T0313736

Thermodynamics

PRINCETON SERIES IN APPLIED MATHEMATICS

Edited by

Ingrid Daubechies, *Princeton University*
Weinan E, *Princeton University*
Jan Karel Lenstra, *Eindhoven University*
Endre Süli, *University of Oxford*

The Princeton Series in Applied Mathematics publishes high quality advanced texts and monographs in all areas of applied mathematics. Books include those of a theoretical and general nature as well as those dealing with the mathematics of specific applications areas and real-world situations.

Thermodynamics

A Dynamical Systems Approach

Wassim M. Haddad

VijaySekhar Chellaboina

Sergey G. Nersesov

PRINCETON UNIVERSITY PRESS

PRINCETON AND OXFORD

In the United Kingdom: Princeton University Press, 3 Market Place, Woodstock,
Oxfordshire OX20 1SY

Library of Congress Cataloging-in-Publication Data

Haddad, Wassim M., 1961–
    Thermodynamics : a dynamical systems approach / Wassim M. Haddad, Vi-
jaySekhar Chellaboina, and Sergey G. Nersesov.
        p. cm. — (Princeton series in applied mathematics)
    Includes bibliographical references and index.
    ISBN: 0-691-12327-6 (acid-free paper)
    1. Thermodynamics—Mathematics. 2. Differentiable dynamical systems. I.
Chellaboina, VijaySekhar, 1970– II. Nersesov, Sergey G., 1976– III. Title. IV.
Series.

QC311.2.H33 2005
536′.7—dc22   2004066029

British Library Cataloging-in-Publication Data is available

This book has been composed in Times Roman in LaTeX

The publisher would like to acknowledge the authors of this volume for providing
the camera-ready copy from which this book was printed.

Printed on acid-free paper. ∞

pup.princeton.edu

Printed in the United States of America

10 9 8 7 6 5 4 3 2 1

To my mother Sofia who made it possible for me to pursue my passion for science, and my wife Lydia who provides the equipoise between this passion and the other joys of life.

W. M. H.

To my children SriHarsha and Saankhya, the entropy agents of my life.

V. C.

To my parents Garry and Ekatherina and my brother Artyom.

S. G. N.

Τα πάντα ρεί. Ποταμείς τοίς αυτοίς εμβαίνομεν τε καί ουκ
εμβαίνομεν, είμεν τε καί ουκ είμεν.

—Herakleitos

[Thermodynamics] is the only physical theory of a universal na-
ture of which I am convinced that it will never be overthrown.

—Albert Einstein

The law that entropy always increases—the second law of
thermodynamics—holds, I think, the supreme position among the
laws of Nature.

—Sir Arthur Eddington

The future belongs to those who can manipulate entropy; those
who understand but energy will be only accountants.

—Frederic Keffer

The energy of the Universe is constant. The entropy of the Uni-
verse tends to a maximum. The total state of the Universe will
inevitably approach a limiting state.

—Rudolf Clausius

Time flows on, never comes back. When the physicist is con-
fronted with this fact he is greatly disturbed.

—Leon Brillouin

The world has signed a pact with the devil; it had to. It is a
covenant to which everything, even every hydrogen atom, is bound.
The terms are clear: if you want to live, you have to die. The
world came into being with the signing of this contract. A scien-
tist calls it the Second Law of Thermodynamics.

—Annie Dillard

Contents

Preface

Thermodynamics is a physical branch of science that governs the thermal behavior of dynamical systems from those as simple as refrigerators to those as complex as our expanding universe. The laws of thermodynamics involving conservation of energy and nonconservation of entropy are, without a doubt, two of the most useful and general laws in all sciences. The first law of thermodynamics, according to which energy cannot be created or destroyed, merely transformed from one form to another, and the second law of thermodynamics, according to which the *usable* energy in an adiabatically isolated dynamical system is always diminishing in spite of the fact that energy is conserved, have had an impact far beyond science and engineering. The second law of thermodynamics is intimately connected to the irreversibility of dynamical processes. In particular, the second law asserts that a dynamical system undergoing a transformation from one state to another cannot be restored to its original state and at the same time restore its environment to its original condition. That is, the status quo cannot be restored everywhere. This gives rise to an increasing quantity known as *entropy*.

Entropy permeates the whole of nature, and unlike energy, which describes the state of a dynamical system, entropy is a measure of change in the status quo of a dynamical system. Hence, the law that entropy always increases, the second law of thermodynamics, defines the direction of time flow and shows that a dynamical system state will continually change in that direction and thus inevitably approach a limiting state corresponding to a state of maximum entropy. It is precisely this irreversibility of all dynamical processes connoting the running down and eventual demise of the universe that has led writers, historians, philosophers, and theologians to ask profound questions such as: How is it possible for life to come into being in a universe governed by a supreme law that impedes the very existence of life?

Even though thermodynamics has provided the foundation for speculation about some of science's most puzzling questions concerning the beginning and the end of the universe, the development of thermodynamics grew out of steam tables and the desire to design and build efficient heat engines, with many scientists and mathematicians expressing concerns about the completeness and clarity of its mathematical foundation over its long and tortuous history. Indeed, many formulations of classical thermodynamics, especially most textbook presentations, poorly amalgamate physics with rigorous mathematics and have had a hard time in finding a balance between nineteenth century steam and heat engine engineering, and twenty first century science and mathematics. In fact, no other discipline in mathematical science is riddled with so many logical and mathematical inconsistencies, differences in definitions, and ill-defined notation as classical thermodynamics. With a notable few exceptions, more than a century of mathematicians have turned away in disquietude from classical thermodynamics, often overlooking its grandiose unsubstantiated claims and allowing it to slip into an abyss of ambiguity.

The development of the theory of thermodynamics followed two conceptually rather different lines of thought. The first (historically), known as *classical thermodynamics*, is based on fundamental laws that are assumed as axioms, which in turn are based on experimental evidence. Conclusions are subsequently drawn from them using the notion of a *thermodynamic state* of a system, which includes temperature, volume, and pressure, among others. The second, known as *statistical thermodynamics*, has its foundation in classical mechanics. However, since the state of a dynamical system in mechanics is completely specified point-wise in time by each point-mass position and velocity and since thermodynamic systems contain large numbers of particles (atoms or molecules, typically on the order of 10^{23}), an ensemble average of different configurations of molecular motion is considered as the state of the system. In this case, the equivalence between heat and dynamical energy is based on a kinetic theory interpretation reducing all thermal behavior to the statistical motions of atoms and molecules. In addition, the second law of thermodynamics has only statistical certainty wherein entropy is directly related to the relative probability of various states of a collection of molecules.

In this monograph, we utilize the *language* of modern mathematics within a theorem-proof format to develop a general dynamical systems theory for reversible and irreversible (non)equilibrium thermodynamics. The monograph is written from a system-theoretic point of view and can be viewed as a contribution to the fields of thermodynamics and mathematical system theory. In particular, we develop a novel formulation of thermodynamics using a middle-ground theory involving deterministic large-scale dynamical system models that bridges the gap between classical and statistical thermodynamics. The benefits of such a theory include the advantage of being independent of the simplifying assumptions that are often made in statistical mechanics and at the same time providing a thermodynamic framework with enough detail of how the system really evolves without ever needing to resort to statistical (subjective or informational) probabilities. In particular, we develop a system-theoretic foundation for thermodynamics using a large-scale dynamical systems perspective. Specifically, using compartmental dynamical system energy flow models, we place the universal energy conservation, energy equipartition, temperature equipartition, and entropy nonconservation laws of thermodynamics on a system-theoretic foundation.

Next, we establish the existence of a *new* and dual notion to entropy, namely, *ectropy*, as a measure of the tendency of a dynamical system to do useful work and grow more organized, and we show that conservation of energy in an adiabatically isolated thermodynamic system necessarily leads to nonconservation of ectropy and entropy. In addition, using the system ectropy as a Lyapunov function candidate, we show that our large-scale thermodynamic energy flow model has convergent trajectories to Lyapunov stable equilibria determined by the large-scale system initial subsystem energies. Furthermore, using the system entropy and ectropy functions, we establish a clear connection between irreversibility, the second law of thermodynamics, and the arrow of time. Finally, these results are generalized to continuum thermodynamics involving infinite-dimensional energy flow conservation models. Since in this case the resulting dynamical system is defined on an infinite-dimensional Banach space that is not locally compact, stability, convergence, and energy equipartition are shown using Sobolev embedding theorems and the notion of generalized (or weak) solutions.

The quest in pursuing a systems foundation for thermodynamics by the authors can be traced back to the late 1980's, when the first author was introduced to the possibility of such a notion by David C. Hyland. David Hyland maintained that the second law of thermodynamics applied to large-scale coupled mechanical systems with modal energy playing the role of temperature, and he asserted that many concepts of energy flow modeling in high-dimensional mechanical systems have clear connections with statistical mechanics of many particle systems. Ensuing discussions between the first author and Dennis S. Bernstein on the thermodynamic behavior of coupled mechanical systems followed periodically over the next several years. However, a key insight in the pursuit of a system-theoretic foundation of thermodynamics came after the first author read Jan C. Willems' seminal two-part paper on dissipative dynamical systems, wherein the second law of thermodynamics is viewed as an axiom in the context of dissipative dynamical systems. The final piece of the puzzle fell into place when the authors started their recent collaboration with James M. Bailey on biological and physiological systems involving energy flow compartmental models, as well as their own collaborative work on vector dissipativity theory of large-scale dynamical systems. The culmination of the above series of research themes has led to this monograph.

The first author would like to thank Dennis S. Bernstein and David C. Hyland for their valuable discussions on thermodynamics and vibrational systems over the years. Their legacy to the first author has been the reinforcement that dynamical system and control theory is exactly at the crossroads between mathematics and engineering, and is in a unique position to clarify and add precision to numerous engineering concepts that are otherwise based on physical insight and empirical observation. The authors would like to thank Tomohisa Hayakawa for carefully reading several versions of the manuscript and providing helpful suggestions. In addition, the authors thank Jan C. Willems for his constructive comments and feedback. Jan Willems' foundational work on dissipative dynamical systems has been a tremendous source of inspiration to the authors. Finally, the authors would like to thank Paul Katinas for providing a translation of Herakleitos' profound statements quoted in ancient Greek on page vi. In addition, we thank Paul for several interesting discussions on the ramifications of these statements to ontology and cosmology.

The results reported in this monograph were obtained at the School of Aerospace Engineering, Georgia Institute of Technology, Atlanta, and the Department of Mechanical and Aerospace Engineering, University of Missouri, Columbia, between June 2003 and August 2004. The research support provided by the Air Force Office of Scientific Research and the National Science Foundation over the years has been instrumental in allowing us to explore basic research topics that have led to some of the material in this monograph. We are indebted to them for their support.

Atlanta, Georgia, USA, March 2005, *Wassim M. Haddad*
Knoxville, Tennessee, USA, March 2005, *VijaySekhar Chellaboina*
Atlanta, Georgia, USA, March 2005, *Sergey G. Nersesov*

Chapter One

Introduction

1.1 An Overview of Thermodynamics

Energy is a concept that underlies our understanding of all physical phenomena and is a measure of the ability of a dynamical system to produce changes (motion) in its own system state as well as changes in the system states of its surroundings. Thermodynamics is a physical branch of science that deals with laws governing energy flow from one body to another and energy transformations from one form to another. These energy flow laws are captured by the fundamental principles known as the first and second laws of thermodynamics. The first law of thermodynamics gives a precise formulation of the equivalence between heat and work and states that among all system transformations, the net system energy is conserved. Hence, energy cannot be created out of nothing and cannot be destroyed; it can merely be transferred from one form to another. The law of conservation of energy is not a mathematical truth, but rather the consequence of an immeasurable culmination of observations over the chronicle of our civilization and is a fundamental *axiom* of the science of heat. The first law does not tell us whether any particular process can actually occur, that is, it does not restrict the ability to convert work into heat or heat into work, except that energy must be conserved in the process. The second law of thermodynamics asserts that while the system energy is always conserved, it will be degraded to a point where it cannot produce any useful work. Hence, it is impossible to extract work from heat without at the same time discarding some heat, giving rise to an increasing quantity known as *entropy*.

While energy describes the state of a dynamical system, entropy refers to changes in the status quo of the system and is a measure of molecular disorder and the amount of wasted energy in a dynamical (energy) transformation from one state (form) to another. Since the system entropy increases, the entropy of a dynamical system tends to a maximum, and thus time, as determined by system entropy increase [70, 89, 105], flows on in one direction only. Even though entropy is a physical property of matter that is not directly observable, it permeates the whole of nature, regulating the *arrow of time*, and is responsible for the enfeeblement and eventual demise of the universe.[1,2] While the laws of thermodynamics form the foundation to basic engineering systems as well as nuclear explosions, cosmology, and our expanding universe, many mathematicians and scientists have expressed concerns about the completeness and clarity of the different expositions of thermodynamics over its long and flexuous history; see [19, 23, 32, 41, 45, 77, 94, 96, 99].

Since the specific motion of every molecule of a thermodynamic system is impossible to predict, a *macroscopic* model of the system is typically used, with appropriate macroscopic states that include pressure, volume, temperature, internal energy, and entropy, among others. One of the key criticisms of the macroscopic viewpoint of thermodynamics, known as *classical thermodynamics*, is the inability of the model to provide enough detail of how the system really evolves; that is, it is lacking a kinetic mechanism for describing the behavior of heat. In developing a kinetic model for heat and dynamical energy, a thermodynamically consistent energy flow model should ensure that the system energy can be modeled by a diffusion (conservation) equation in the form of a *parabolic* partial differential equation. Such systems are infinite-dimensional, and hence, finite-dimensional approximations are of very high order, giving rise to large-scale dynamical systems. Since energy is a fundamental concept in the analysis of large-scale dynamical systems, and heat (energy) is a fundamental concept of thermodynamics involving the capacity of hot bodies (more energetic subsystems) to produce work, thermodynamics is a theory of large-scale dynamical systems.

High-dimensional dynamical systems can arise from both macroscopic and *microscopic* points of view. Microscopic thermodynamic

---

[1]Many natural philosophers have associated this ravaging irrecoverability in connection to the second law of thermodynamics with an eschatological terminus of the universe. Namely, the creation of a certain degree of life and order in the universe is inevitably coupled with an even greater degree of death and disorder. A convincing derivation of this bold claim has, however, never been given.

[2]The earliest perception of irreversibility of nature and the universe along with time's arrow was postulated by the ancient Greek philosopher Herakleitos ($\sim$ 535– $\sim$ 475 B.C.). Herakleitos' profound statements (quoted in ancient Greek on page vi), *Everything is in a state of flux* and *Man cannot step into the same river twice, because neither the man nor the river is the same*, created the foundation for all other speculation on physics and metaphysics. The idea that the universe is in constant change and that there is an underlying order to this change—the *Logos*, dare we say—postulates the very existence of entropy as a physical property of matter permeating the whole of nature and the universe.

models can have the form of a distributed-parameter model or a large-scale system model comprised of a large number of interconnected subsystems. In contrast to macroscopic models involving the evolution of global quantities (e.g., energy, temperature, entropy, etc.), microscopic models are based upon the modeling of local quantities that describe the atoms and molecules that make up the system and their speeds, energies, masses, angular momenta, behavior during collisions, etc. The mathematical formulations based on these quantities form the basis of *statistical mechanics*. Thermodynamics based on statistical mechanics is known as *statistical thermodynamics* and involves the mechanics of an ensemble of many particles (atoms or molecules) wherein the detailed description of the system state loses importance and only average properties of large numbers of particles are considered. Since microscopic details are obscured on the macroscopic level, it is appropriate to view a microscopic model as an inherent model of uncertainty. However, for a thermodynamic system the macroscopic and microscopic quantities are related since they are simply different ways of describing the same phenomena. Thus, if the global macroscopic quantities can be expressed in terms of the local microscopic quantities, the laws of thermodynamics could be described in the language of statistical mechanics. This interweaving of the microscopic and macroscopic points of view leads to diffusion being a natural consequence of dimensionality and, hence, uncertainty on the microscopic level, despite the fact that there is no uncertainty about the diffusion process per se.

Thermodynamics was spawned from the desire to design and build efficient heat engines, and it quickly spread to speculations about the universe upon the discovery of entropy as a fundamental physical property of matter. The theory of classical thermodynamics was predominantly developed by Carnot, Clausius, Kelvin, Planck, Gibbs, and Carathéodory,[3] and its laws have become one of the most firmly established scientific achievements ever accomplished. The pioneering work of Carnot [24] was the first to establish the impossibility of a *perpetuum mobile* of the second kind[4] by constructing a cyclical process

---

[3]The theory of classical thermodynamics has also been developed over the last one and a half centuries by many other researchers. Notable contributions include the work of Rankine, Reech, Clapeyron, and Giles.

[4]A *perpetuum mobile* of the second kind is a cyclic device that would continuously extract heat from the environment and completely convert it into mechanical work. Since such a machine would not create energy, it would not violate the first law of thermodynamics. In contrast, a machine that creates its own energy and thus violates the first law is called a *perpetuum mobile* of the first kind.

(now known as the Carnot cycle) involving two competing cycles, and showing that it is impossible to extract work from heat without at the same time discarding some heat. Carnot's main assumption (now known as Carnot's principle) was that it is impossible to perform an arbitrarily often repeatable cycle whose only effect is to produce an unlimited amount of positive work. In particular, Carnot showed that the *efficiency* of a reversible cycle—that is, the ratio of the total work produced during the cycle and the amount of heat transferred from a boiler (furnace) to a cooler (refrigerator)—is bounded by a universal maximum, and this maximum is only a function of the temperatures of the boiler and the cooler. Both heat reservoirs (i.e., furnace and refrigerator) are assumed to have an infinite source of heat so that their state is unchanged by their heat exchange with the engine (i.e., the device that performs the cycle), and hence, the engine is capable of repeating the cycle arbitrarily often. Carnot's result was remarkably arrived at using the erroneous concept that heat is an indestructible substance, that is, the *caloric theory of heat*.[5]

Using a macroscopic approach and building on the work of Carnot, Clausius [26–29] was the first to introduce the notion of entropy as a physical property of matter and to establish the two main laws of thermodynamics involving conservation of energy[6] and nonconservation of entropy. Specifically, using conservation of energy principles, Clausius showed that Carnot's principle is valid. Furthermore, Clausius postulated that it is impossible to perform a cyclic system transformation whose only effect is to transfer heat from a body at a given temperature to a body at a higher temperature. From this postulate Clausius established the second law of thermodynamics as a statement about entropy increase for *adiabatically isolated systems* (i.e., systems with no heat exchange with the environment). From this statement Clausius goes on to state what have become known as the most controversial words in the history of thermodynamics and perhaps all of science; namely, the entropy of the universe is tending to a maximum, and the total state of the universe will inevitably ap-

---

[5]After Carnot's death, several articles were discovered wherein he had expressed doubt about the caloric theory of heat (i.e., the conservation of heat). However, these articles were not published until the late 1870's, and as such, did not influence Clausius in rejecting the caloric theory of heat and deriving Carnot's results using the energy equivalence principle of Mayer and Joule.

[6]Even though many scientists are credited with the law of conservation of energy, it was first discovered independently by Mayer and Joule, with Joule providing a series of decisive, quantitative studies in the 1840's showing the equivalence between heat and mechanical work.

proach a limiting state. The fact that the entropy of the universe is a thermodynamically undefined concept led to serious criticism of Clausius' grand universal generalizations by many of his contemporaries as well as numerous scientists, natural philosophers, and theologians who followed.[7] In his later work [29], Clausius remitted his famous claim that the entropy of the universe is tending to a maximum.

In parallel research Kelvin [55, 93] developed similar, and in some cases identical, results as Clausius, with the main difference being the absence of the concept of entropy. Kelvin's main view of thermodynamics was that of a universal irreversibility of physical phenomena occurring in nature. Kelvin further postulated that it is impossible to perform a cyclic system transformation whose only effect is to transform into work heat from a source that is at the same temperature throughout.[8] Without any supporting mathematical arguments, Kelvin goes on to state that the universe is heading towards a state of eternal rest wherein all life on Earth in the distant future shall perish. This claim by Kelvin involving a universal tendency towards dissipation has come to be known as the *heat death of the universe.*

Building on the work of Clausius and Kelvin, Planck [82, 83] refined the formulation of classical thermodynamics. From 1897 to 1964, Planck's treatise [82] underwent eleven editions. Nevertheless, these editions have several inconsistencies regarding key notions and definitions of reversible and irreversible processes. Planck's main theme of thermodynamics is that entropy increase is a necessary and sufficient condition for irreversibility. Without any proof (mathematical or otherwise), he goes on to conclude that every dynamical system in nature

---

[7]Clausius' concept of the universe approaching a limiting state was inadvertently based on an analogy between a universe and a finite adiabatically isolated system possessing a finite energy content. It is not clear where the heat absorbed by the system, if that system is the universe, needed to define the change in entropy between two system states comes from. Nor is it clear whether an infinite and endlessly expanding universe governed by the theory of general relativity has a final equilibrium state. An additional caveat is the delineation of energy conservation when changes in the curvature of space-time need to be accounted for. In this case, the energy density tensor in Einstein's equations is only covariantly conserved since it does not account for gravitational energy—an unsolved problem in the general theory of relativity. Nevertheless, the law of conservation of energy is as close to an absolute truth as our incomprehensible universe will allow us to deduce.

[8]In the case of thermodynamic systems with positive absolute temperatures, Kelvin's postulate can be shown to be equivalent to Clausius' postulate. However, many textbooks erroneously show this equivalence without the assumption of *positive* absolute temperatures. Physical systems possessing a small number of energy levels with negative absolute temperatures are discussed in [71, 86].

evolves in such a way that the total entropy of all of its parts increases. In the case of reversible processes, he concludes that the total entropy remains constant. Unlike Clausius' entropy increase conclusion, Planck's increase entropy principle is not restricted to adiabatically isolated dynamical systems. Rather, it applies to all system transformations wherein the initial states of any exogenous system, belonging to the environment and coupled to the transformed dynamical system, return to their initial condition.

Unlike the work of Clausius, Kelvin, and Planck involving cyclical system transformations, the work of Gibbs [39] involves system equilibrium states. Specifically, Gibbs assumes a thermodynamic state of a system involving pressure, volume, temperature, energy, and entropy, among others, and proposes that an *isolated system*[9] (i.e., a system with no energy exchange with the environment) is in equilibrium if and only if all possible variations of the state of the system that do not alter its energy, the variation of the system entropy is negative semidefinite. Hence, Gibbs' principle gives necessary and sufficient conditions for a thermodynamically stable equilibrium and should be viewed as a variational principle defining admissible (i.e., stable) equilibrium states. Thus, it does not provide any information about the dynamical state of the system as a function of time nor any conclusion regarding entropy increase in a dynamical system transformation.

Carathéodory [20, 21] was the first to give a rigorous axiomatic mathematical framework for thermodynamics. In particular, using an equilibrium thermodynamic theory, Carathéodory assumes a state space endowed with a Euclidean topology and defines the equilibrium state of the system using thermal and deformation coordinates. Next, he defines an *adiabatic accessibility* relation wherein a reachability condition of an adiabatic process[10] is used such that an empirical statement of the second law characterizes a mathematical structure for an abstract state space. Carathéodory's postulate for the second law states that in every open neighborhood of any state of a system, there exist states such that for some second open neighborhood contained in the first neighborhood, all the states in the second neighborhood cannot be reached by adiabatic processes from states in the first neighborhood. From this postulate Carathéodory goes on to show

---

[9]Gibbs' principle is weaker than Clausius' principle leading to the second law involving entropy increase since it holds for the more restrictive case of isolated systems.

[10]Carathéodory's definition of an adiabatic process is nonstandard and involves transformations that take place while the system remains in an *adiabatic container*. For details see [20, 21].

that for a special class of systems, which he called *simple systems*, there exists a *locally* defined entropy and an absolute temperature on the state space for every simple system equilibrium state. One of the key weaknesses of Carathéodory's work is that his principle is too weak in establishing the existence of a *global* entropy function.

Adopting a microscopic viewpoint, Boltzmann [15] was the first to give a probabilistic interpretation of entropy involving different configurations of molecular motion of the microscopic dynamics. Specifically, Boltzmann reinterpreted thermodynamics in terms of molecules and atoms by relating the *mechanical* behavior of individual atoms with their *thermodynamic* behavior by suitably averaging properties of the individual atoms. In particular, even though individual atoms are assumed to obey the laws of Newtonian mechanics, by suitably averaging over the velocity distribution of these atoms Boltzmann showed how the microscopic (mechanical) behavior of atoms and molecules produced effects visible on a macroscopic (thermodynamic) scale. He goes on to argue that Clausius' thermodynamic entropy (a macroscopic quantity) is proportional to the logarithm of the probability that a system will exist in the state it is in relative to all possible states it could be in. Thus, the entropy of a thermodynamic system state (macrostate) corresponds to the degree of uncertainty about the actual system mechanical state (microstate) when only the thermodynamic system state (macrostate) is known. Hence, the essence of Boltzmann thermodynamics is that thermodynamic systems with a constant energy level will evolve from a less probable state to a more probable state with the equilibrium system state corresponding to a state of maximum entropy (i.e., highest probability).

In the first half of the twentieth century, the macroscopic and microscopic interpretations of thermodynamics underwent a long and fierce debate. To exacerbate matters, since classical thermodynamics was formulated as a physical theory and not a mathematical theory, many scientists and mathematical physicists expressed concerns about the completeness and clarity of the mathematical foundation of thermodynamics [5,19,96]. In fact, many fundamental conclusions arrived at by classical thermodynamics can be viewed as paradoxical. For example, in classical thermodynamics the notion of entropy (and temperature) is only defined for equilibrium states. However, the theory concludes that nonequilibrium states transition towards equilibrium states as a consequence of the law of entropy increase! Furthermore, classical thermodynamics is mainly restricted to systems in equilibrium. The second law infers that for any transformation occurring in an isolated system, the entropy of the final state can never be less than the entropy of the initial state. In this context, the initial and final states of the system are equilibrium states. However, by definition, an equilibrium state is a system state that has the property that whenever the state of the system starts at the equilibrium state it will remain at the equilibrium state for all future time unless an external disturbance acts on the system. Hence, the entropy of the system can only increase if the system is *not* isolated! Many aspects of classical thermodynamics are riddled with such inconsistencies, and hence it is not surprising that many formulations of thermodynamics, especially most textbook expositions, poorly amalgamate physics with rigorous mathematics. Perhaps this is best eulogized in [96, p. 6], wherein Truesdell describes the present state of the theory of thermodynamics as a "dismal swamp of obscurity." More recently, Arnold [5, p. 163] writes that "every mathematician knows it is impossible to understand an elementary course in thermodynamics."

As we have outlined, it is clear that there have been many different presentations of classical thermodynamics with varying hypotheses and conclusions. To exacerbate matters, the careless and considerable differences in the definitions of two of the key notions of thermodynamics—namely, the notions of reversibility and irreversibility—have contributed to the widespread confusion and lack of clarity of the exposition of classical thermodynamics over the past one and a half centuries. For example, the concept of reversible processes as defined by Clausius, Kelvin, Planck, and Carathéodory have very different meanings. In particular, Clausius defines a reversible (*umkehrbar*) process as a slowly varying process wherein successive states of this process differ by infinitesimals from the equilibrium system states. Such system transformations are commonly referred to as *quasistatic* transformations in the thermodynamic literature. Alternatively, Kelvin's notions of reversibility involve the ability of a system to completely recover its initial state from the final system state. Planck introduced several notions of reversibility. His main notion of reversibility is one of *complete* reversibility and involves recoverability of the original state of the dynamical system while at the same time restoring the environment to its original condition. Unlike Clausius' notion of reversibility, Kelvin's and Planck's notions of reversibility do not require the system to exactly retrace its original trajectory in reverse order. Carathéodory's notion of reversibility involves recoverability of the system state in an adiabatic process resulting in yet another definition of thermodynamic reversibility. These subtle distinctions of (ir)reversibility are often unrecognized in the thermodynamic literature. Notable exceptions to this fact include [16, 97], with [97] providing an excellent exposition of the relation between irreversibility, the second law of thermodynamics, and the arrow of time.

The arrow of time[11] remains one of physics' most perplexing enigmas [36,43,53,60,69,83,87,102]. Even though time is one of the most familiar concepts humankind has ever encountered, it is the least understood. Puzzling questions of time's mysteries have remained unanswered throughout the centuries. Questions such as, Where does time come from? What would our universe look like without time? Can there be more than one dimension to time? Is time truly a fundamental appurtenance woven into the fabric of the universe, or is it just a useful edifice for organizing our perception of events? Why is the concept of time hardly ever found in the most fundamental physical laws of nature and the universe? Can we go back in time? And if so, can we change past events?

Human experience perceives time flow as unidirectional; the present is forever flowing towards the future and away from a forever fixed past. Many scientists have attributed this *emergence* of the direction of time flow to the second law of thermodynamics due to its intimate connection to the irreversibility of dynamical processes.[12] In this regard, thermodynamics is disjoint from Newtonian and Hamiltonian mechanics (including Einstein's extensions), since these theories are invariant under time reversal, that is, they make no distinction between one direction of time and the other. Such theories possess a *time-reversal symmetry*, wherein, from any given moment of time, the governing laws treat past and future in exactly the same way [61]. For example, a film run backwards of a harmonic oscillator over a full period or a planet orbiting the Sun would represent possible events. In contrast, a film run backwards of water in a glass coalescing into a solid ice cube or ashes self-assembling into a log of wood would immediately be identified as an impossible event. The idea that the second law of thermodynamics provides a physical foundation for the arrow

---

[11]Perhaps a better expression here is the *geodesic arrow of time*, since, as Einstein's theory of relativity shows, time and space are intricately coupled, and hence one cannot curve space without involving time as well. Thus, time has a shape that goes along with its directionality.

[12]In statistical thermodynamics the arrow of time is viewed as a consequence of high system dimensionality and randomness. However, since in statistical thermodynamics it is not absolutely certain that entropy increases in every dynamical process, the direction of time, as determined by entropy increase, has only statistical certainty and not an absolute certainty. Hence, it cannot be concluded from statistical thermodynamics that time has a unique direction of flow.

of time has been postulated by many authors [37, 85, 87]. However, a convincing argument of this claim has never been given [43, 60, 97].

In the last half of the twentieth century, thermodynamics was reformulated as a global nonlinear field theory with the ultimate objective to determine the independent field variables of this theory [31, 75, 95]. This aspect of thermodynamics, which became known as *rational thermodynamics*, was predicated on an entirely new axiomatic approach. As a result of this approach, modern continuum thermodynamics was developed using theories from elastic materials, viscous materials, and materials with memory [30, 34, 35, 44]. The main difference between classical thermodynamics and rational thermodynamics can be traced back to the fact that in rational thermodynamics the second law is not interpreted as a restriction on the transformations a system can undergo, but rather as a restriction on the system's constitutive equations.

More recently, a major contribution to equilibrium thermodynamics is given in [65]. This work builds on the work of Carathéodory [20, 21] and Giles [40] by developing a thermodynamic system representation involving a state space on which an adiabatic accessibility relation is defined. The existence and uniqueness of an entropy function is established as a consequence of adiabatic accessibility among equilibrium states. As in Carathéodory's work, the authors in [65] also restrict their attention to simple (possibly interconnected) systems in order to arrive at an entropy increase principle. However, it should be noted that the notion of a simple system in [65] is not equivalent to that of Carathéodory's notion of a simple system. Connections between thermodynamics and system theory as well as information theory have also been explored in the literature [9, 10, 17, 18, 48, 80, 100, 102, 104, 106]. For an excellent exposition of these different facets of thermodynamics see [42].

Thermodynamic principles have also been repeatedly used in coupled mechanical systems to arrive at energy flow models with modal energy playing the role of temperature. Specifically, in an attempt to approximate high-dimensional dynamics of large-scale structural (oscillatory) systems with a low-dimensional diffusive (non-oscillatory) dynamical model, structural dynamicists have developed thermodynamic energy flow models using stochastic energy flow techniques. In particular, statistical energy analysis (SEA) predicated on averaging system states over the statistics of the uncertain system parameters has been extensively developed for mechanical and acoustic vibration problems [22, 54, 62, 69, 91, 103]. Thermodynamic models are derived from large-scale dynamical systems of discrete subsystems involving stored energy flow among subsystems based on the assumption of weak subsystem coupling or identical subsystems. However, the ability of SEA to predict the dynamic behavior of a complex large-scale dynamical system in terms of pairwise subsystem interactions is severely limited by the coupling strength of the remaining subsystems on the subsystem pair. Hence, it is not surprising that SEA energy flow predictions for large-scale systems with strong coupling can be erroneous.

Alternatively, a deterministic thermodynamically motivated energy flow modeling for structural systems is addressed in [57–59]. This approach exploits energy flow models in terms of thermodynamic energy (i.e., the ability to dissipate heat) as opposed to stored energy and is not limited to weak subsystem coupling. A stochastic energy flow *compartmental model* (i.e., a model characterized by energy conservation laws) predicated on averaging system states over the statistics of stochastic system exogenous disturbances is developed in [10]. The basic result demonstrates how linear compartmental models arise from second-moment analysis of state space systems under the assumption of weak coupling. Even though these results can be potentially applicable to linear large-scale dynamical systems with weak coupling, such connections are not explored in [10]. With the notable exception of [22], none of the aforementioned SEA-related works addresses the second law of thermodynamics involving entropy notions in the energy flow between subsystems.

1.2 System Thermodynamics

In contrast to mechanics, which is based on a dynamical system theory, classical thermodynamics is a physical theory and does not possess equations of motion. The goal of the present monograph is directed towards placing thermodynamics on a system-theoretic foundation so as to harmonize it with classical mechanics. In particular, we develop a novel formulation of thermodynamics that can be viewed as a moderate-sized system theory as compared to statistical thermodynamics. This middle-ground theory involves deterministic large-scale dynamical system models that bridge the gap between classical and statistical thermodynamics. Specifically, since thermodynamic models are concerned with energy flow among subsystems, we use a state space formulation to develop a nonlinear compartmental dynamical system model that is characterized by energy conservation laws capturing the exchange of energy between coupled macroscopic subsystems. Furthermore, using graph-theoretic notions, we state two thermodynamic axioms consistent with the zeroth and second laws of thermodynamics, which ensure that our large-scale dynamical system model gives rise to a thermodynamically consistent energy flow model. Specifically, using a large-scale dynamical systems theory perspective for thermodynamics, we show that our compartmental dynamical system model leads to a precise formulation of the equivalence between work energy and heat in a large-scale dynamical system.

Since our thermodynamic formulation is based on a large-scale dynamical system theory involving the exchange of energy with conservation laws describing transfer, accumulation, and dissipation between subsystems and the environment, our framework goes beyond classical thermodynamics characterized by a purely empirical theory, wherein a physical system is viewed as an input-output *black box* system. Furthermore, unlike classical thermodynamics, which is often limited to the description of systems in equilibrium states, our approach addresses nonequilibrium thermodynamic systems. This allows us to connect and unify the behavior of heat as described by the equations of thermal transfer and as described by classical thermodynamics. This exposition demonstrates that these disciplines of classical physics are derivable from the same principles and are part of the same mathematical framework.

Our nonequilibrium thermodynamic framework goes beyond the reciprocal relations for irreversible processes developed by Onsager[13] [78, 79] and further extended by Casimir [25], which fall short of a complete dynamical theory. The Onsager-Casimir reciprocal relations treat only the irreversible aspects of system processes, and thus the theory is an algebraic theory that is primarily restricted to describing (time-independent) system steady states. In addition, the Onsager-Casimir formalism is restricted to linear systems, wherein a linearity restriction is placed on the admissible constitutive relations between the thermodynamic forces and fluxes. Another limitation of the Onsager-Casimir framework is the difficulty in providing a macroscopic description for large-scale complex dynamical systems. In contrast, the proposed system thermodynamic formalism brings classical thermodynamics within the framework of modern nonlinear dynamical systems theory, thus providing information about the dynamical behavior between the initial and final equilibrium system states.

---

[13]Onsager's theorem pertains to the thermodynamics of linear systems, wherein a symmetric reciprocal relation applies between forces and fluxes. In particular, a flow or flux of matter in thermodiffusion is caused by the force exerted by the thermal gradient. Conversely, a concentration gradient causes a heat flow, an effect that has been experimentally verified.

Next, we give a deterministic definition of entropy for a large-scale dynamical system that is consistent with the classical thermodynamic definition of entropy, and we show that it satisfies a Clausius-type inequality leading to the law of entropy nonconservation. However, unlike classical thermodynamics, wherein entropy is not defined for arbitrary states out of equilibrium, our definition of entropy holds for nonequilibrium dynamical systems. Furthermore, we introduce a *new* and dual notion to entropy—namely, *ectropy*[14]—as a measure of the tendency of a large-scale dynamical system to do useful work and grow more organized, and we show that conservation of energy in an adiabatically isolated thermodynamically consistent system necessarily leads to nonconservation of ectropy and entropy. Hence, for every dynamical transformation in an adiabatically isolated thermodynamically consistent system, the entropy of the final system state is greater than or equal to the entropy of the initial system state. Then, using the system ectropy as a Lyapunov function candidate, we show that in the absence of energy exchange with the environment our thermodynamically consistent large-scale nonlinear dynamical system model possesses a continuum of equilibria and is *semistable*, that is, it has convergent subsystem energies to Lyapunov stable energy equilibria determined by the large-scale system initial subsystem energies. In addition, we show that the steady-state distribution of the large-scale system energies is uniform, leading to system energy equipartitioning corresponding to a minimum ectropy and a maximum entropy equilibrium state.

For our thermodynamically consistent dynamical system model, we further establish the existence of a *unique* continuously differentiable global entropy and ectropy function for all equilibrium and nonequilibrium states. Using these global entropy and ectropy functions, we go on to establish a clear connection between thermodynamics and the arrow of time. Specifically, we rigorously show a *state irrecoverability* and hence a *state irreversibility*[15] nature of thermodynamics.

---

[14]Ectropy comes from the Greek word $\varepsilon\kappa\tau\rho o\pi\eta$ ($\varepsilon\kappa$ and $\tau\rho o\pi\eta$) for outward transformation and is the literal antonym of entropy ($\varepsilon\nu\tau\rho o\pi\eta$—$\varepsilon\nu$ and $\tau\rho o\pi\eta$), signifying an inward transformation. The word *entropy* was proposed by Clausius for its phonetic similarity to energy with the additional connotation reflecting change ($\tau\rho o\pi\eta$) and does not necessarily correspond to an inward or outward change.

[15]In the terminology of [97], state irreversibility is referred to as *time-reversal non-invariance*. However, since the term *time-reversal* is not meant literally (that is, we consider dynamical systems whose trajectory reversal is or is not allowed and *not* a reversal of time itself), state reversibility is a more appropriate expression.

In particular, we show that for every nonequilibrium system state and corresponding system trajectory of our thermodynamically consistent large-scale nonlinear dynamical system, there does not exist a state such that the corresponding system trajectory completely recovers the initial system state of the dynamical system and at the same time restores the energy supplied by the environment back to its original condition. This, along with the existence of a global strictly increasing entropy function on every nontrivial system trajectory, gives a clear *time-reversal asymmetry* characterization of thermodynamics, establishing an emergence of the direction of time flow. In the case where the subsystem energies are proportional to subsystem temperatures, we show that our dynamical system model leads to temperature equipartition, wherein all the system energy is transferred into heat at a uniform temperature. Furthermore, we show that our system-theoretic definition of entropy and the newly proposed notion of ectropy are consistent with Boltzmann's kinetic theory of gases involving an n-body theory of ideal gases divided by diathermal walls. Finally, these results are generalized to continuum thermodynamics involving infinite-dimensional energy flow conservation models.

1.3 A Brief Outline of the Monograph

The objective of this monograph is to develop a system-theoretic foundation for classical thermodynamics using dynamical systems and control notions. The main contents of the monograph are as follows. In Chapter 2, we establish notation and definitions, and we review some basic results on nonnegative and compartmental dynamical systems needed to establish thermodynamically consistent energy flow models. Furthermore, we introduce the notions of (ir)reversible and (ir)recoverable dynamical systems, as well as volume-preserving flows and recurrent dynamical systems. In Chapter 3, we use a large-scale dynamical systems perspective to provide a system-theoretic foundation for thermodynamics. Specifically, using a state space formulation, we develop a nonlinear compartmental dynamical system model characterized by energy conservation laws that is consistent with basic thermodynamic principles. In particular, using the total subsystem energies as a candidate system energy storage function, we show that our thermodynamic system is lossless and hence can deliver to its surroundings all of its stored subsystem energies and can store all of the work done to all of its subsystems. This leads to the first law of thermodynamics involving conservation of energy and places no limitation on the possibility of transforming heat into work or work into heat.

Next, we show that the classical Clausius equality and inequality for reversible and irreversible thermodynamics are satisfied over cyclic motions for our thermodynamically consistent energy flow model and guarantee the existence of a continuous system entropy function. In addition, we establish the existence of a *unique*, continuously differentiable global entropy function for our large-scale dynamical system, which is used to define inverse subsystem temperatures as the derivative of the subsystem entropies with respect to the subsystem energies. Then we turn our attention to stability and convergence. Specifically, using the system ectropy as a Lyapunov function candidate, we show that in the absence of energy exchange with the environment, the proposed thermodynamic model is semistable with a uniform energy distribution corresponding to a state of minimum ectropy and a state of maximum entropy. Furthermore, using the system entropy and ectropy functions, we develop a clear connection between irreversibility, the second law of thermodynamics, and the entropic arrow of time.

In Chapter 4, we generalize the results of Chapter 3 to the case where the subsystem energies in the large-scale dynamical system model are proportional to subsystem temperatures, and we arrive at temperature equipartition for the proposed thermodynamic model. Furthermore, we provide a kinetic theory interpretation of the steady-state expressions for entropy and ectropy. In Chapter 5, we augment our nonlinear compartmental dynamical system model with an additional (deformation) state representing compartmental volumes to arrive at a general statement of the first law of thermodynamics, giving a precise formulation of the equivalence between heat and work. In addition, we use the proposed augmented nonlinear compartmental dynamical system model in conjunction with a Carnot-like cycle analysis to show the equivalence between the classical Kelvin-Planck and Clausius postulates of the second law of thermodynamics. In Chapter 6, we specialize the results of Chapter 3 to thermodynamic systems with linear energy exchange.

In Chapter 7, we extend the results of Chapter 3 to continuum thermodynamic systems, wherein the subsystems are uniformly distributed over an n-dimensional (not necessarily Euclidean) space. Specifically, we develop a nonlinear distributed parameter model wherein the system energy is modeled by a diffusion (conservation) equation in the form of a parabolic partial differential equation. Energy equipartition and semistability are shown using the well-known Sobolev embedding theorems and the notion of generalized (or weak) solutions. This exposition shows that the behavior of heat, as described by the equations of thermal transport and as described by classical thermodynamics, is derivable from the same principles and is part of the same scientific discipline, and thus provides a unification between Fourier's theory of heat conduction and classical thermodynamics. Finally, we draw conclusions in Chapter 8.

Chapter Two

Dynamical System Theory

2.1 Notation, Definitions, and Mathematical Preliminaries

As discussed in Chapter 1, in this monograph we develop thermodynamic system models using large-scale nonlinear compartmental dynamical systems. The mathematical foundation for compartmental modeling is *nonnegative dynamical system theory* [47, 90], which involves dynamical systems with nonnegative state variables. Since our thermodynamic state equations govern energy flow between subsystems, it follows from physical arguments that system energy initial conditions give rise to trajectories that remain in the nonnegative orthant of the state space. In this chapter we introduce notation, several definitions, and some key results on nonlinear nonnegative dynamical systems needed for developing the main results of this monograph.

The notation used in this monograph is fairly standard. Specifically, $\mathbb{R}$ denotes the set of real numbers, $\overline{\mathbb{Z}}_+$ denotes the set of nonnegative integers, $\mathbb{Z}_+$ denotes the set of positive integers, $\mathbb{R}^n$ denotes the set of $n \times 1$ column vectors, $(\cdot)^{\mathrm{T}}$ denotes transpose, $(\cdot)^{\#}$ denotes group generalized inverse, and I_n or I denotes the $n \times n$ identity matrix. For $v \in \mathbb{R}^q$ we write $v \geq\geq 0$ (respectively, $v >> 0$) to indicate that every component of v is nonnegative (respectively, positive). In this case we say that v is *nonnegative* or *positive*, respectively. Let $\overline{\mathbb{R}}^q_+$ and $\mathbb{R}^q_+$ denote the nonnegative and positive orthants of $\mathbb{R}^q$, that is, if $v \in \mathbb{R}^q$, then $v \in \overline{\mathbb{R}}^q_+$ and $v \in \mathbb{R}^q_+$ are equivalent, respectively, to $v \geq\geq 0$ and $v >> 0$. Furthermore, we denote the boundary, the interior, and the closure of the set $\mathcal{S}$ by $\partial\mathcal{S}$, $\overset{\circ}{\mathcal{S}}$, and $\overline{\mathcal{S}}$, respectively.

We write $\|\cdot\|$ for the Euclidean vector norm, $\|\cdot\|_{\mathcal{B}}$ for the operator norm of an element in a Banach space $\mathcal{B}$, $\mathcal{R}(M)$ and $\mathcal{N}(M)$ for the range space and the null space of a matrix M, $\mathrm{spec}(M)$ for the spectrum of the square matrix M, and $\mathrm{ind}(M)$ for the index of M (that is, the size of the largest Jordan block of M associated with $\lambda = 0$, where $\lambda \in \mathrm{spec}(M)$). Furthermore, we write $V'(x)$ for the Fréchet derivative of V at x, $\mathcal{B}_\varepsilon(\alpha)$, $\alpha \in \mathbb{R}^n$, $\varepsilon > 0$, for the open ball centered at α with radius ε, $M \geq 0$ (respectively, $M > 0$) to denote the fact that the Hermitian matrix M is nonnegative (respectively, positive) definite, inf to denote infimum (that is, the greatest lower bound), sup to denote supremum (that is, the least upper bound), and $x(t) \to \mathcal{M}$ as $t \to \infty$ to denote that $x(t)$ approaches the set $\mathcal{M}$ (that is, for each $\varepsilon > 0$ there exists $T > 0$ such that $\mathrm{dist}(x(t), \mathcal{M}) < \varepsilon$ for all $t > T$, where $\mathrm{dist}(p, \mathcal{M}) \triangleq \inf_{x \in \mathcal{M}} \|p - x\|_{\mathcal{B}}$). Finally, the notions of openness, convergence, continuity, and compactness that we use throughout the monograph refer to the topology generated on $\overline{\mathbb{R}}_+^q$ (respectively, $\mathcal{B}$) by the norm $\| \cdot \|$ (respectively, $\| \cdot \|_{\mathcal{B}}$).

The following definition introduces the notion of Z-, M-, essentially nonnegative, compartmental, and nonnegative matrices.

Definition 2.1 ([6, 10, 48]) *Let* $W \in \mathbb{R}^{q \times q}$. W *is a* Z-*matrix if* $W_{(i,j)} \le 0$, $i, j = 1, \ldots, q$, $i \ne j$, *where* $W_{(i,j)}$ *denotes the* (i, j)th *entry of* W. W *is an* M-*matrix (respectively, a* nonsingular M-*matrix) if* W *is a* Z-*matrix and all the principal minors of* W *are nonnegative (respectively, positive).* W *is* essentially nonnegative *if* $-W$ *is a* Z-*matrix, that is,* $W_{(i,j)} \ge 0$, $i, j = 1, \ldots, q$, $i \ne j$. W *is* compartmental *if* W *is essentially nonnegative and* $\sum_{i=1}^{q} W_{(i,j)} \le 0$, $j = 1, \ldots, q$. *Finally,* W *is* nonnegative[1] *(respectively,* positive*) if* $W_{(i,j)} \ge 0$ *(respectively,* $W_{(i,j)} > 0$), $i, j = 1, \ldots, q$.

The following definition introduces the notion of essentially nonnegative functions [8, 47].

Definition 2.2 *Let* $w = [w_1, \ldots, w_q]^{\mathrm{T}} : \mathcal{V} \to \mathbb{R}^q$, *where* $\mathcal{V}$ *is an open subset of* $\mathbb{R}^q$ *that contains* $\overline{\mathbb{R}}_+^q$. *Then* w *is* essentially nonnegative *if* $w_i(z) \ge 0$ *for all* $i = 1, \ldots, q$ *and* $z \in \overline{\mathbb{R}}_+^q$ *such that* $z_i = 0$, *where* z_i *denotes the* ith *component of* z.

Note that if $w(z) = Wz$, where $W \in \mathbb{R}^{q \times q}$, then $w(\cdot)$ is essentially nonnegative if and only if W is an essentially nonnegative matrix. For the following result we consider the nonlinear dynamical system

$$\dot{z}(t) = w(z(t)), \quad z(t_0) = z_0, \quad t \in \mathcal{I}_{z_0}, \tag{2.1}$$

where $z(t) \in \mathcal{D} \subseteq \mathbb{R}^q$, $t \in \mathcal{I}_{z_0}$, is the system state vector, $\mathcal{D}$ is an open subset of $\mathbb{R}^q$, $w : \mathcal{D} \to \mathbb{R}^q$ is locally Lipschitz continuous on $\mathcal{D}$, and $\mathcal{I}_{z_0} = [t_0, \tau_{z_0})$, $t_0 < \tau_{z_0} \le \infty$, is the maximal interval of existence for the solution $z(\cdot)$ of (2.1). A function $z : \mathcal{I}_{z_0} \to \mathcal{D}$ is said to be a

---

[1]In this monograph it is important to distinguish between a square nonnegative (respectively, positive) matrix and a nonnegative-definite (respectively, positive-definite) matrix.

solution to (2.1) on the interval $\mathcal{I}_{z_0} \subseteq [t_0, \infty)$ with *initial condition* $z(t_0) = z_0$ if and only if $z(t)$ satisfies (2.1) for all $t \in \mathcal{I}_{z_0}$. Recall that the point $z_e \in \mathcal{D}$ is an *equilibrium point* of (2.1) if and only if $w(z_e) = 0$. Furthermore, a subset $\mathcal{D}_c \subseteq \mathcal{D}$ is an *invariant set* with respect to (2.1) if and only if $z(t_0) \in \mathcal{D}_c$ implies that $z(t) \in \mathcal{D}_c$ for all $t \in \mathcal{I}_{z_0} \subset \mathbb{R}$. Finally, recall that if all solutions to (2.1) are bounded, then it follows from the Peano-Cauchy theorem [50, pp. 16, 17] that $\mathcal{I}_{z_0} = \mathbb{R}$.

Proposition 2.1 *Suppose* $\overline{\mathbb{R}}_+^q \subset \mathcal{V}$. *Then* $\overline{\mathbb{R}}_+^q$ *is an invariant set with respect to (2.1) if and only if* $w : \mathcal{V} \to \mathbb{R}^q$ *is essentially nonnegative.*

Proof. Define $\text{dist}(z, \overline{\mathbb{R}}_+^q) \triangleq \inf_{y \in \overline{\mathbb{R}}_+^q} \|z - y\|$, $z \in \mathbb{R}^q$. Now, suppose $w : \mathcal{V} \to \mathbb{R}^q$ is essentially nonnegative and let $z \in \overline{\mathbb{R}}_+^q$. For every $i \in \{1, \dots, q\}$ such that $z_i = 0$, it follows that $z_i + hw_i(z) = hw_i(z) \geq 0$ for $h \geq 0$, while for every $i \in \{1, \dots, q\}$ such that $z_i > 0$, $z_i + hw_i(z) > 0$ for sufficiently small $|h|$. Thus, $z + hw(z) \in \overline{\mathbb{R}}_+^q$ for all sufficiently small $h > 0$, and hence $\lim_{h \to 0^+} \text{dist}(z + hw(z), \overline{\mathbb{R}}_+^q)/h = 0$. It now follows from Theorem 4.1.28 of [1], with $z(t_0) = z$, that $z(t) \in \overline{\mathbb{R}}_+^q$ for all $t \in \mathcal{I}_{z_0}$.

Conversely, suppose $z(t_0) = z \in \overline{\mathbb{R}}_+^q$ and assume, *ad absurdum*, that there exists $i \in \{1, \dots, q\}$ such that $z_i = 0$ and $w_i(z) < 0$. Then it follows from continuity that there exists sufficiently small $h > 0$ such that $w_i(z(t)) < 0$ for all $t \in [t_0, t_0 + h)$. Hence, $z_i(t)$ is strictly decreasing, and thus, $z(t) \notin \overline{\mathbb{R}}_+^q$ for all $t \in (t_0, t_0 + h)$, which leads to a contradiction. $\qquad\square$

If $w(z) = Wz$, where $W \in \mathbb{R}^{q \times q}$, the solution to (2.1) is standard and is given by $z(t) = e^{Wt} z(0)$, $t \geq 0$. The following corollary to Proposition 2.1 is immediate.

Corollary 2.1 *Let* $W \in \mathbb{R}^{q \times q}$. *Then* W *is essentially nonnegative if and only if* e^{Wt} *is nonnegative for all* $t \geq 0$.

Proof. The proof is a direct consequence of Proposition 2.1 with $w(z) = Wz$. The proof can also be shown using basic matrix mathematics [10]. Specifically, if W is essentially nonnegative, then there exists sufficiently large $\alpha > 0$ such that $W_\alpha \triangleq W + \alpha I$ is nonnegative. Hence, $e^{W_\alpha t} = e^{(W + \alpha I)t} \geq\geq 0$, $t \geq 0$, and thus $e^{Wt} = e^{-\alpha t} e^{W_\alpha t} \geq\geq 0$, $t \geq 0$.

Conversely, suppose $e^{Wt} \geq\geq 0$, $t \geq 0$, and assume, *ad absurdum*, there exist i, j such that $i \neq j$ and $W_{(i,j)} < 0$. Now, since $e^{Wt} =$

$\sum_{k=0}^{\infty}(k!)^{-1}W^k t^k$, it follows that

$$[e^{Wt}]_{(i,j)} = I_{(i,j)} + tW_{(i,j)} + \mathcal{O}(t^2). \qquad (2.2)$$

Thus, as $t \to 0$ and $i \neq j$, it follows that $[e^{Wt}]_{(i,j)} < 0$ for some t sufficiently small, which leads to a contradiction. Hence, W is essentially nonnegative. $\square$

It follows from Proposition 2.1 that if $z_0 \geq\geq 0$, then $z(t) \geq\geq 0$, $t \geq t_0$, if and only if $w(\cdot)$ is essentially nonnegative. In this case, we say that (2.1) is a *nonnegative dynamical system*.

2.2 Stability Theory for Nonnegative Dynamical Systems

One of the main contributions of this monograph is the connection of ectropy and entropy to the stability properties of thermodynamic processes. System stability is characterized by analyzing the response of a dynamical system to small perturbations in the system states. Specifically, an equilibrium point of a dynamical system is said to be *stable* or *Lyapunov stable* if, for small values of initial disturbances, the perturbed system motion remains in an arbitrarily prescribed small region of the state space. More precisely, stability is equivalent to continuity of solutions as a function of the system initial conditions over a neighborhood of the equilibrium point uniformly in time. If, in addition, all solutions of the dynamical system approach the equilibrium point for large values of time, then the equilibrium point is said to be *asymptotically stable*. In this monograph we apply Lyapunov methods [14,56] to show that our thermodynamically consistent large-scale dynamical system is *semistable*, that is, system trajectories converge to Lyapunov stable equilibrium states that depend upon the system initial conditions.[2]

In this section, we establish key stability results for nonlinear nonnegative dynamical systems. Since the proofs of these results are virtually identical to the standard stability proofs for nonlinear dynamical systems, they are not presented here. In particular, standard Lyapunov stability theorems and invariant set theorems for nonlinear dynamical systems [56] can be used directly for nonnegative dynamical systems with the required sufficient conditions verified on $\overline{\mathbb{R}}_+^q$.

---

[2]The concept of semistability involves a stability notion that lies perfectly between Lyapunov stability and asymptotic stability, and is vital in addressing the stability of systems having a continuum of equilibria [12,13].

The following definition introduces several types of stability for the nonnegative dynamical system (2.1) with $\mathcal{I}_{z_0} = [t_0, \infty)$.

Definition 2.3 *The equilibrium solution $z(t) \equiv z_e$ of (2.1) is Lyapunov stable if, for every $\varepsilon > 0$, there exists $\delta = \delta(\varepsilon) > 0$ such that if $z_0 \in \mathcal{B}_\delta(z_e) \cap \overline{\mathbb{R}}_+^q$, then $z(t) \in \mathcal{B}_\varepsilon(z_e) \cap \overline{\mathbb{R}}_+^q$, $t \geq t_0$. The equilibrium solution $z(t) \equiv z_e$ of (2.1) is semistable if it is Lyapunov stable and there exists $\delta > 0$ such that if $z_0 \in \mathcal{B}_\delta(z_e) \cap \overline{\mathbb{R}}_+^q$, then $\lim_{t \to \infty} z(t)$ exists and corresponds to a Lyapunov stable equilibrium point. The equilibrium solution $z(t) \equiv z_e$ of (2.1) is asymptotically stable if it is Lyapunov stable and there exists $\delta > 0$ such that if $z_0 \in \mathcal{B}_\delta(z_e) \cap \overline{\mathbb{R}}_+^q$, then $\lim_{t \to \infty} z(t) = z_e$. Finally, the equilibrium solution $z(t) \equiv z_e$ of (2.1) is globally asymptotically stable if the previous statement holds for all $z_0 \in \overline{\mathbb{R}}_+^q$.*

Recall that a square matrix $W \in \mathbb{R}^{q \times q}$ is *semistable* if and only if $\lim_{t \to \infty} e^{Wt}$ exists [10, 47], while W is *asymptotically stable* if and only if $\lim_{t \to \infty} e^{Wt} = 0$. The following result, known as Lyapunov's direct method, gives sufficient conditions for Lyapunov and asymptotic stability of a nonlinear nonnegative dynamical system. For this result let $V : \mathcal{D} \to \mathbb{R}$ be a continuously differentiable function with derivative *along the trajectories* of (2.1) given by $\dot{V}(z) \triangleq V'(z)w(z)$. Note that $\dot{V}(z)$ is dependent on the system dynamics (2.1). Since, using the chain rule,

$$\dot{V}(z) = \frac{\mathrm{d}}{\mathrm{d}t} V(s(t, z)) \Big|_{t=t_0} = V'(z)w(z), \qquad (2.3)$$

where $s(\cdot, z_0)$ denotes the solution to (2.1) with initial condition $z(t_0) = z_0$, it follows that if $\dot{V}(z)$ is negative, then $V(z)$ decreases along the solutions $s(t, z)$ of (2.1) through $z \in \mathcal{D}$ at $t = t_0$.

Theorem 2.1 *Let $\mathcal{D}$ be an open subset of $\mathbb{R}^q$ that contains $\overline{\mathbb{R}}_+^q$. Consider the nonlinear nonnegative dynamical system (2.1) and assume that there exists a continuously differentiable function $V : \mathcal{D} \to \mathbb{R}$ such that*

$$V(z_e) = 0, \qquad (2.4)$$
$$V(z) > 0, \qquad z \in \mathcal{D}, \qquad z \neq z_e, \qquad (2.5)$$
$$V'(z)w(z) \leq 0, \qquad z \in \mathcal{D}, \qquad (2.6)$$

where $z_e \in \overline{\mathbb{R}}_+^q$ is an equilibrium point of (2.1). Then the equilibrium solution $z(t) \equiv z_e$ to (2.1) is Lyapunov stable. If, in addition,

$$V'(z)w(z) < 0, \qquad z \in \mathcal{D}, \qquad z \neq z_e, \qquad (2.7)$$

then the equilibrium solution $z(t) \equiv z_e$ to (2.1) is asymptotically stable.

   A continuously differentiable function $V(\cdot)$ satisfying (2.4) and (2.5) is called a *Lyapunov function candidate* for the nonlinear nonnegative dynamical system (2.1). If, additionally, $V(\cdot)$ satisfies (2.6), then $V(\cdot)$ is called a *Lyapunov function* for the nonlinear nonnegative dynamical system (2.1). In light of the conditions (2.4)–(2.7), $V(\cdot)$ can be regarded as a generalized *energy function* for the nonlinear nonnegative dynamical system (2.1).

   Next, we state a global asymptotic stability theorem for (2.1). First, however, recall that a function $V : \overline{\mathbb{R}}_+^q \to \mathbb{R}$ satisfying

$$V(z) \to \infty \quad \text{as} \quad \|z\| \to \infty \tag{2.8}$$

is called *proper* or *radially unbounded*.

Theorem 2.2 *Consider the nonlinear nonnegative dynamical system (2.1) and assume there exists a continuously differentiable function $V : \overline{\mathbb{R}}_+^q \to \mathbb{R}$ such that*

$$V(z_e) = 0, \tag{2.9}$$

$$V(z) > 0, \qquad z \in \overline{\mathbb{R}}_+^q, \qquad z \neq z_e, \tag{2.10}$$

$$V'(z)w(z) < 0, \qquad z \in \overline{\mathbb{R}}_+^q, \qquad z \neq z_e, \tag{2.11}$$

$$V(z) \to \infty \quad as \quad \|z\| \to \infty, \tag{2.12}$$

where $z_e \in \overline{\mathbb{R}}_+^q$ is an equilibrium point of (2.1). Then the equilibrium solution $z(t) \equiv z_e$ to (2.1) is globally asymptotically stable.

   Note that the radial unboundedness condition (2.12) assures that the constant *energy surfaces*, or *level sets*, $V(z) = \alpha$, $\alpha > 0$, are closed curves, and hence, since the system trajectories move from one energy surface to an inner energy surface, the system trajectories cannot drift away from the system equilibrium. The Lyapunov function is thus decreasing with the dynamics of the system, that is, the trajectories of the dynamical system cut the level sets of the Lyapunov function.

   Next, we introduce the Krasovskii-LaSalle invariant set theorem to relax one of the conditions on the Lyapunov function $V(\cdot)$ in Theorems 2.1 and 2.2. In particular, the strict negative definiteness condition on the Lyapunov derivative can be relaxed while assuring system asymptotic stability. Specifically, if a continuously differentiable function defined on a compact invariant set (in $\mathcal{D}$) with respect to the nonlinear nonnegative dynamical system (2.1) can be constructed whose derivative along the system's trajectories is negative semidefinite and no system trajectories can stay indefinitely at points where the function's derivative vanishes, then the system's equilibrium point is asymptotically stable. This result follows from the Krasovskii-LaSalle invariant set theorem for nonlinear nonnegative dynamical systems, which we now state.

Theorem 2.3 *Consider the nonlinear nonnegative dynamical system (2.1), assume $\mathcal{D}_c \subset \mathcal{D}$ is a compact invariant set with respect to (2.1), and assume there exists a continuously differentiable function $V \colon \mathcal{D}_c \to \mathbb{R}$ such that $V'(z)w(z) \leq 0$, $z \in \mathcal{D}_c$. Let $\mathcal{R} \triangleq \{z \in \mathcal{D}_c : V'(z)w(z) = 0\}$, and let $\mathcal{M}$ be the largest invariant set contained in $\mathcal{R}$. If $z(t_0) \in \mathcal{D}_c$, then $z(t) \to \mathcal{M}$ as $t \to \infty$.*

The next result is a direct consequence of Theorem 2.3 and does not require the existence of a compact invariant set $\mathcal{D}_c \subset \mathcal{D}$ with respect to (2.1). In particular, if the solutions to (2.1) are bounded for all initial conditions in $\overline{\mathbb{R}}_+^q$, then, with $\mathcal{D}_c = \overline{\mathbb{R}}_+^q$, it follows from Theorem 2.3 that attraction to $\mathcal{M}$ is guaranteed globally.

Corollary 2.2 *Consider the nonlinear nonnegative dynamical system (2.1), assume that the solutions to (2.1) are bounded for all initial conditions $z(t_0) \in \overline{\mathbb{R}}_+^q$, and assume there exists a continuously differentiable function $V \colon \overline{\mathbb{R}}_+^q \to \mathbb{R}$ such that $V'(z)w(z) \leq 0$, $z \in \overline{\mathbb{R}}_+^q$. Let $\mathcal{R} \triangleq \{z \in \overline{\mathbb{R}}_+^q : V'(z)w(z) = 0\}$, and let $\mathcal{M}$ be the largest invariant set contained in $\mathcal{R}$. If $z(t_0) \in \overline{\mathbb{R}}_+^q$, then $z(t) \to \mathcal{M}$ as $t \to \infty$.*

Next, using Theorem 2.3 we provide a generalization of Theorem 2.1 for local asymptotic stability of a nonlinear nonnegative dynamical system. For this result, recall that if the equilibrium solution $z(t) \equiv z_e$ to (2.1) is asymptotically stable, then the *domain of attraction* $\mathcal{D}_A \subseteq \mathcal{D}$ of (2.1) is given by

$$\mathcal{D}_A \triangleq \{z_0 \in \mathcal{D} : \text{if } z(t_0) = z_0, \text{ then } \lim_{t \to \infty} z(t) = z_e\}. \quad (2.13)$$

Corollary 2.3 *Consider the nonlinear nonnegative dynamical system (2.1), assume $\mathcal{D}_c \subset \mathcal{D}$ is a compact invariant set with respect to (2.1) such that $z_e \in \mathcal{D}_c$ is an equilibrium point of (2.1), and assume that there exists a continuously differentiable function $V \colon \mathcal{D}_c \to \mathbb{R}$ such that $V(z_e) = 0$, $V(z) > 0$, $z \neq z_e$, and $V'(z)w(z) \leq 0$, $z \in \mathcal{D}_c$. Furthermore, assume that the set $\mathcal{R} \triangleq \{z \in \mathcal{D}_c : V'(z)w(z) = 0\}$ contains no invariant set other than the set $\{z_e\}$. Then the equilibrium solution $z(t) \equiv z_e$ to (2.1) is asymptotically stable, and $\mathcal{D}_c$ is a subset of the domain of attraction of (2.1).*

Finally, we present a global invariant set theorem for guarantee-ing global asymptotic stability of a nonlinear nonnegative dynamical system.

Theorem 2.4 *Consider the nonlinear nonnegative dynamical system (2.1) and assume there exists a continuously differentiable function* $V : \overline{\mathbb{R}}_+^q \to \mathbb{R}$ *such that*

$$V(z_e) = 0, \tag{2.14}$$

$$V(z) > 0, \quad z \in \overline{\mathbb{R}}_+^q, \quad z \neq z_e, \tag{2.15}$$

$$V'(z)w(z) \leq 0, \quad z \in \overline{\mathbb{R}}_+^q, \tag{2.16}$$

$$V(z) \to \infty \ as \ \|z\| \to \infty, \tag{2.17}$$

where $z_e \in \overline{\mathbb{R}}_+^q$ *is an equilibrium point of (2.1). Furthermore, assume that the set* $\mathcal{R} \triangleq \{z \in \overline{\mathbb{R}}_+^q : V'(z)w(z) = 0\}$ *contains no invariant set other than the set* $\{z_e\}$*. Then the equilibrium solution* $z(t) \equiv z_e$ *to (2.1) is globally asymptotically stable.*

In the remainder of this section, we discuss dynamical systems de-fined on Banach spaces and present some of the key results on in-variant set stability theorems for infinite-dimensional dynamical sys-tems. For infinite-dimensional systems, Lyapunov stability theorems are similar to the Lyapunov theorems just presented with the required conditions verified on the Banach space $\mathcal{B}$ [14]. However, since norms are not equivalent in infinite-dimensional spaces, Lyapunov, semi-, and asymptotic stability are obtained with respect to a specific norm $\|\cdot\|_{\mathcal{B}}$ defined on $\mathcal{B}$.

Definition 2.4 *Let* $\mathcal{B}$ *be a Banach space with norm* $\|\cdot\|_{\mathcal{B}}$*. A dynam-ical system on* $\mathcal{B}$ *is the triple* $(\mathcal{B}, [t_0, \infty), s)$*, where* $s : [t_0, \infty) \times \mathcal{B} \to \mathcal{B}$ *is such that the following axioms hold:*

  i) (Continuity): $s(\cdot, \cdot)$ *is jointly continuous.*

  ii) (Consistency): $s(t_0, z_0) = z_0$ *for all* $t_0 \in \mathbb{R}$ *and* $z_0 \in \mathcal{B}$*.*

  iii) (Semigroup property): $s(t + \tau, z_0) = s(\tau, s(t, z_0))$ *for all* $z_0 \in \mathcal{B}$ *and* $t, \tau \in [t_0, \infty)$*.*

The above definition of a dynamical system can be generalized to in-clude external system disturbances $u(\cdot) \in \mathcal{U}$, where $\mathcal{U}$ is an input space consisting of bounded continuous $U \subseteq \mathbb{R}^m$-valued functions on $[t_0, \infty)$. In this case, the dynamical system on $\mathcal{B}$ is defined by the pentuple $(\mathcal{B},$

$\mathcal{U}$, $[t_0, \infty)$, s, h), where $s : [t_0, \infty) \times \mathcal{B} \times \mathcal{U} \to \mathcal{B}$ and $h : \mathcal{B} \times U \to Y$ defines a memoryless *read-out map* $y(t) = h(s(t, z_0, u), u(t))$ for all $z_0 \in \mathcal{B}$, $u(\cdot) \in \mathcal{U}$, and $t \in [t_0, \infty)$. Here, $y(\cdot) \in \mathcal{Y}$, where $\mathcal{Y}$ denotes an output space and $y(t)$ belongs to the fixed set $Y \subseteq \mathbb{R}^l$ for all $t \geq t_0$. In this case, to assure causality of the dynamical system one needs to invoke an additional axiom (determinism axiom), which assures that the state, and hence the output, of the dynamical system before some time τ are not influenced by the values of the output after time τ. For further details see Definition 2.7.

Henceforth, we denote the dynamical system $(\mathcal{B}, [t_0, \infty), s)$ by $\mathcal{G}$ and we refer to the map $s(\cdot, \cdot)$ as the *flow* or *trajectory* of $\mathcal{G}$ corresponding to $z_0 \in \mathcal{B}$, and for a given $s(t, z_0)$, $t \geq t_0$, we refer to $z_0 \in \mathcal{B}$ as an *initial condition* of $\mathcal{G}$. Given $t \in \mathbb{R}$ we denote the map $s(t, \cdot) : \mathcal{B} \to \mathcal{B}$ by $s_t(z_0)$. Hence, for a fixed $t \in \mathbb{R}$ the set of mappings defined by $s_t(z_0) = s(t, z_0)$ for every $z_0 \in \mathcal{B}$ gives the *flow* of $\mathcal{G}$. In particular, if $\mathcal{B}_0$ is a collection of initial conditions such that $\mathcal{B}_0 \subset \mathcal{B}$, then the flow $s_t : \mathcal{B}_0 \to \mathcal{B}$ is the motion of all points $z_0 \in \mathcal{B}_0$ or, equivalently, the image of $\mathcal{B}_0 \subset \mathcal{B}$ under the flow s_t, that is, $s_t(\mathcal{B}_0) \subset \mathcal{B}$, where $s_t(\mathcal{B}_0) \triangleq \{y : y = s_t(z_0) \text{ for all } z_0 \in \mathcal{B}_0\}$. Alternatively, if the initial condition $z_0 \in \mathcal{B}$ is fixed and we let $[t_0, t_1] \subset \mathbb{R}$, then the mapping $s(\cdot, z_0) : [t_0, t_1] \to \mathcal{B}$ defines the *solution curve* or *trajectory* of the dynamical system $\mathcal{G}$. Hence, the mapping $s(\cdot, z_0)$ generates a graph in $[t_0, t_1] \times \mathcal{B}$ identifying the trajectory corresponding to the motion along a curve through the point z_0 in a subset $\mathcal{B}$ of the state space. Given $z \in \mathcal{B}$, we denote the map $s(\cdot, z) : \mathbb{R} \to \mathcal{B}$ by $s^z(t)$. Finally, we define a *positive orbit* through the point $z_0 \in \mathcal{B}$ as the motion along the curve

$$\mathcal{O}_{z_0}^+ \triangleq \{z \in \mathcal{B} : z = s(t, z_0), \ t \geq t_0\}. \tag{2.18}$$

Definition 2.5 *Let $\mathcal{G}$ be a dynamical system defined on $\mathcal{B}$. A point $p \in \mathcal{B}$ is a* positive limit point *of the trajectory $s(\cdot, z)$ if there exists a monotonic sequence $\{t_n\}_{n=0}^{\infty}$ of nonnegative numbers, with $t_n \to \infty$ as $n \to \infty$, such that $s(t_n, z) \to p$ as $n \to \infty$. The set of all positive limit points of $s(t, z)$, $t \geq t_0$, is the* positive limit set *$\omega(z)$ of $\mathcal{G}$.*

In the mathematical literature, the positive limit set is often referred to as the *ω-limit set*. Note that if $p \in \mathcal{B}$ is a positive limit point of the trajectory $s(\cdot, z)$, then for all $\varepsilon > 0$ and finite time $T > 0$ there exists $t > T$ such that $\|s(t, z) - p\|_\mathcal{B} < \varepsilon$. This follows from the fact that $\|s(t, z) - p\|_\mathcal{B} < \varepsilon$ for all $\varepsilon > 0$, and some $t > T > 0$ is equivalent to the existence of a sequence $\{t_n\}_{n=0}^{\infty}$, with $t_n \to \infty$ as $n \to \infty$, such that $s(t_n, z) \to p$ as $n \to \infty$.

Definition 2.6 *A set* $\mathcal{M} \subset \mathcal{B}$ *is a* positively invariant set *with respect to the dynamical system* $\mathcal{G}$ *if* $s_t(\mathcal{M}) \subseteq \mathcal{M}$ *for all* $t \geq t_0$*, where* $s_t(\mathcal{M}) \triangleq \{s_t(z) : z \in \mathcal{M}\}$ *and* $s_t(z) \triangleq s(t, z)$*,* $z \in \mathcal{B}$*,* $t \geq t_0$*. A set* $\mathcal{M} \subseteq \mathcal{B}$ *is an* invariant set *with respect to the dynamical system* $\mathcal{G}$ *if* $s_t(\mathcal{M}) = \mathcal{M}$ *for all* $t \in [t_0, \infty)$*.*

Next, we state a key proposition involving positive limit sets for the infinite-dimensional dynamical system $\mathcal{G}$.

Proposition 2.2 ([49]) *Let* $\mathcal{G}$ *be a dynamical system defined on* $\mathcal{B}$ *and suppose that the positive orbit* $\mathcal{O}_z^+$ *through* z *of* $\mathcal{G}$ *belongs to a compact subset of* $\mathcal{B}$*. Then the positive limit set* $\omega(z)$ *of* $\mathcal{O}_z^+$ *is a nonempty, compact, connected invariant set.*

The following result presents an extension of the Krasovskii-LaSalle invariant set theorem (Theorem 2.3) to infinite-dimensional dynamical systems. This result holds for *undisturbed* dynamical systems (i.e., $u(t) \equiv 0$), as well as for disturbed dynamical systems wherein the input space consists of one constant element only, that is, $u(t) \equiv u^*$. For the statement of this result, define

$$\dot{V}(z) \triangleq \lim_{h \to 0^+} \frac{1}{h}[V(s(t_0 + h, z)) - V(z)], \quad z \in \mathcal{B}, \qquad (2.19)$$

for a given continuous function $V : \mathcal{B} \to \mathbb{R}$ and for every $z \in \mathcal{B}$ such that the limit in (2.19) exists.

Theorem 2.5 ([49]) *Consider a dynamical system* $\mathcal{G}$ *defined on a Banach space* $\mathcal{B}$*. Let* $\mathcal{B}_c \subset \mathcal{B}$ *be a closed set, and assume there exists a continuous function* $V : \mathcal{B}_c \to \mathbb{R}$ *such that* $\dot{V}(z) \leq 0$*,* $z \in \mathcal{B}_c$*. Furthermore, let* $\mathcal{R} \triangleq \{z \in \mathcal{B}_c : \dot{V}(z) = 0\}$*, and let* $\mathcal{M}$ *denote the largest invariant set (with respect to the dynamical system* $\mathcal{G}$*) contained in* $\mathcal{R}$*. Then for every* $z_0 \in \mathcal{B}_c$ *such that* $\mathcal{O}_{z_0}^+ \subset \mathcal{B}_c$ *and* $\mathcal{O}_{z_0}^+$ *is contained in a compact subset of* $\mathcal{B}$*,* $s(t, z_0) \to \mathcal{M}$ *as* $t \to \infty$*.*

In order to apply Theorem 2.5, one needs to show that the positive orbit $\mathcal{O}_{z_0}^+$ of $\mathcal{G}$ is contained in a compact subset of $\mathcal{B}$. Even though for finite-dimensional systems this is a direct consequence of boundedness of solutions, for infinite-dimensional systems local boundedness of an orbit of $\mathcal{G}$ does not ensure that the orbit belongs to a compact subset of $\mathcal{B}$. In light of this, we have the following result. For the statement of this result, let $\mathcal{B}$ and $\mathcal{C}$ be Banach spaces and recall that $\mathcal{B}$ is *compactly embedded* in $\mathcal{C}$ if $\mathcal{B} \subset \mathcal{C}$ and a unit ball in $\mathcal{B}$ belongs to a compact subset in $\mathcal{C}$.

Theorem 2.6 ([49]) *Let $\mathcal{B}$ and $\mathcal{C}$ be Banach spaces such that $\mathcal{B}$ is compactly embedded in $\mathcal{C}$, and let $\mathcal{G}$ be a dynamical system defined on $\mathcal{B}$ and $\mathcal{C}$. Assume there exist continuous functions $V_{\mathcal{B}} : \mathcal{B} \to \mathbb{R}$ and $V_{\mathcal{C}} : \mathcal{C} \to \mathbb{R}$ such that $\dot{V}_{\mathcal{B}}(z) \leq 0$, $z \in \mathcal{B}_{\mathrm{c}}$, and $\dot{V}_{\mathcal{C}}(z) \leq 0$, $z \in \mathcal{C}_{\mathrm{c}}$, where $\mathcal{B}_{\mathrm{c}} = \{z \in \mathcal{B} : V_{\mathcal{B}}(z) < \eta\}$ and $\mathcal{C}_{\mathrm{c}} = \{z \in \mathcal{C} : V_{\mathcal{C}}(z) < \eta\}$ for some $\eta > 0$ such that $\mathcal{B}_{\mathrm{c}} \subset \mathcal{C}_{\mathrm{c}}$. If $\mathcal{B}_{\mathrm{c}}$ is bounded, then for every $z_0 \in \mathcal{B}_{\mathrm{c}}$, $s(t, z_0) \to \mathcal{M}$ in $\mathcal{C}$ as $t \to \infty$, where $\mathcal{M}$ denotes the largest invariant set contained in $\mathcal{R}$ given by*

$$\mathcal{R} = \{z \in \overline{\mathcal{C}_{\mathrm{c}}} : \dot{V}_{\mathcal{C}}(z) = 0\}. \tag{2.20}$$

Theorem 2.6 can also be used to establish existence (in t) of (generalized) solutions of infinite-dimensional dynamical systems $\mathcal{G}$ over the semi-infinite interval $[t_0, \infty)$. In particular, global existence can be obtained by constructing a Lyapunov function $V : \mathcal{B} \to \mathbb{R}$ and invoking the continuation Peano-Cauchy theorem to obtain a dynamical system $\mathcal{G}$ of a subset $\mathcal{B}$ of the space $\mathcal{C}$. For further details see [49].

2.3 Reversibility, Irreversibility, Recoverability, and Irrecoverability

The notions of reversibility, irreversibility, recoverability, and irrecoverability all play a crucial role in thermodynamic processes. In this section we define the notions of *R-state reversibility*, *state reversibility*, and *state recoverability* of a dynamical system $\mathcal{G}$. *R*-state reversibility concerns the existence of a system state with the property that a transformed system trajectory through an involution operator R is an image of a given system trajectory of $\mathcal{G}$ on a specified finite time interval. State reversibility concerns the existence of a system state with the property that the resulting system trajectory is the time-reversed image of a given system trajectory of $\mathcal{G}$ on a specified finite time interval. Finally, state recoverability concerns the existence of a system state with the property that the resulting system trajectory completely recovers the initial state of the dynamical system over a finite time interval.

To establish the notions of (ir)reversibility and (ir)recoverability of a dynamical system $\mathcal{G}$ defined on a Banach space $\mathcal{B}$, we require a generalization of Definition 2.4. For this definition, $\mathcal{U}$ is an input space and consists of bounded continuous U-valued functions on $[0, \infty)$. The set $U \subseteq \mathbb{R}^m$ contains the set of input values, that is, at any time $t \geq t_0$, $u(t) \in U$. The space $\mathcal{U}$ is assumed to be closed under the shift operator, that is, if $u \in \mathcal{U}$, then the function u_T defined by $u_T(t) \triangleq u(t+T)$ is contained in $\mathcal{U}$ for all $T \geq 0$. Furthermore, $\mathcal{Y}$ is an output space and consists of continuous Y-valued functions on $[0, \infty)$. The set $Y \subseteq \mathbb{R}^l$ contains the set of output values, that is, each value of $y(t) \in Y$, $t \geq t_0$. The space $\mathcal{Y}$ is assumed to be closed under the shift operator, that is, if $y \in \mathcal{Y}$, then the function y_T defined by $y_T(t) \triangleq y(t+T)$ is contained in $\mathcal{Y}$ for all $T \geq 0$.

Definition 2.7 *Let $\mathcal{B}$ be a Banach space with norm $\| \cdot \|_{\mathcal{B}}$. A dynamical system on $\mathcal{B}$ is the octuple $(\mathcal{B}, \mathcal{U}, U, \mathcal{Y}, Y, [0, \infty), s, h)$, where $s : [0, \infty) \times \mathcal{B} \times \mathcal{U} \to \mathcal{B}$ and $h : \mathcal{B} \times U \to Y$ are such that the following axioms hold:*

  i) (Continuity): $s(\cdot, \cdot, u)$ is jointly continuous for all $u \in \mathcal{U}$.

  ii) (Consistency): $s(t_0, z_0, u) = z_0$ for all $t_0 \in \mathbb{R}$, $z_0 \in \mathcal{B}$, and $u \in \mathcal{U}$.

  iii) (Determinism): $s(t, z_0, u_1) = s(t, z_0, u_2)$ for all $t \in [t_0, \infty)$, $z_0 \in \mathcal{B}$, and u_1, $u_2 \in \mathcal{U}$ satisfying $u_1(\tau) = u_2(\tau)$, $\tau \leq t$.

  iv) (Semi-group property): $s(\tau, s(t, z_0, u), u) = s(t+\tau, z_0, u)$ for all $z_0 \in \mathcal{B}$, $u \in \mathcal{U}$, and τ, $t \in [t_0, \infty)$.

  v) (Read-out map): There exists $y \in \mathcal{Y}$ such that $y(t) = h(s(t, z_0, u), u(t))$ for all $z_0 \in \mathcal{B}$, $u \in \mathcal{U}$, and $t \geq t_0$.

As in Section 2.2, we denote the dynamical system $(\mathcal{B}, \mathcal{U}, U, \mathcal{Y}, Y, [0, \infty), s, h)$ by $\mathcal{G}$. In general, the output of $\mathcal{G}$ depends on both the present input of $\mathcal{G}$ and the past history of $\mathcal{G}$. Hence, the output at some time t_1 depends on the state $s(t_1, z_0, u)$ of $\mathcal{G}$, which effectively serves as an information storage (memory) of past history. Furthermore, the determinism axiom assures that the state and thus the output before some time t_1 are not influenced by the values of the output after time t_1. Hence, future inputs to $\mathcal{G}$ do not effect past and present outputs of $\mathcal{G}$. This is simply a statement of causality that holds for all physical systems. Finally, we note that the read-out map is memoryless in the sense that outputs only depend on the instantaneous (present) values of the state and input.

For the next set of definitions the following notation is needed. For a given interval $[t_0, t_1]$, where $0 \leq t_0 < t_1 < \infty$, let $\mathcal{W}_{[t_0, t_1]}$ denote the set of all possible trajectories of $\mathcal{G}$ given by

$$\mathcal{W}_{[t_0, t_1]} \triangleq \{s^z : [t_0, t_1] \times \mathcal{U} \to \mathcal{B} : s^z(\cdot, u(\cdot)) \text{ satisfies Axioms } i) - iv)$$
$$\text{of Definition 2.7, } z \in \mathcal{B}, \text{ and } u(\cdot) \in \mathcal{U}\}, \qquad (2.21)$$

where $s^z(\cdot, u(\cdot))$ denotes the solution curve or trajectory of $\mathcal{G}$ for a given fixed initial condition $z \in \mathcal{B}$ and input $u(\cdot) \in \mathcal{U}$.

Definition 2.8 *Consider the dynamical system $\mathcal{G}$ defined on $\mathcal{B}$. Let $R : \mathcal{B} \to \mathcal{B}$ be an involutive operator (that is, $R^2 = I_\mathcal{B}$, where $I_\mathcal{B}$ denotes the identity operator on $\mathcal{B}$) and let $s^z(\cdot, u(\cdot)) \in \mathcal{W}_{[t_0, t_1]}$, where $u(\cdot) \in \mathcal{U}$. The function $s^{-z} : [t_0, t_1] \times \mathcal{U} \to \mathcal{B}$ is an R-reversed trajectory of $s^z(\cdot, u(\cdot))$ if there exists an input $u^-(\cdot) \in \mathcal{U}$ and a continuous, strictly increasing function $\tau : [t_0, t_1] \to [t_0, t_1]$ such that $\tau(t_0) = t_0$, $\tau(t_1) = t_1$, and*

$$s^{-z}(t, u^-(t)) = Rs^z(t_0 + t_1 - \tau(t), u(t_0 + t_1 - \tau(t))), \quad t \in [t_0, t_1].$$
$$(2.22)$$

Definition 2.9 *Consider the dynamical system $\mathcal{G}$ defined on $\mathcal{B}$. Let $R : \mathcal{B} \to \mathcal{B}$ be an involutive operator, let $r : \mathcal{U} \times \mathcal{Y} \to \mathbb{R}$, and let $s^z(\cdot, u(\cdot)) \in \mathcal{W}_{[t_0, t_1]}$, where $u(\cdot) \in \mathcal{U}$. $s^z(\cdot, u(\cdot))$ is an R-reversible trajectory of $\mathcal{G}$ if there exists an input $u^-(\cdot) \in \mathcal{U}$ such that $s^{-z}(\cdot, u^-(\cdot)) \in \mathcal{W}_{[t_0, t_1]}$ and*

$$\int_{t_0}^{t_1} r(u(t), y(t))\mathrm{d}t + \int_{t_0}^{t_1} r(u^-(t), y^-(t))\mathrm{d}t = 0, \qquad (2.23)$$

where $y^-(\cdot)$ denotes the read-out map for the R-reversed trajectory of $s^z(\cdot, u(\cdot))$. Furthermore, $\mathcal{G}$ is an R-state reversible dynamical system if for every $z \in \mathcal{B}$, $s^z(\cdot, u(\cdot))$, where $u(\cdot) \in \mathcal{U}$, is an R-reversible trajectory of $\mathcal{G}$.

In classical mechanics, R is a transformation which reverses the sign of all system momenta, whereas in classical reversible thermodynamics R can be taken to be the identity operator. Note that if $R = I_\mathcal{B}$, then $s^z(\cdot, u(\cdot))$, where $u(\cdot) \in \mathcal{U}$, is an $I_\mathcal{B}$-reversible trajectory or, simply, $s^z(\cdot, u(\cdot))$ is a *reversible trajectory*. Furthermore, we say that $\mathcal{G}$ is a *state reversible dynamical system* if and only if for every $z \in \mathcal{B}$, $s^z(\cdot, u(\cdot))$, where $u(\cdot) \in \mathcal{U}$, is a reversible trajectory of $\mathcal{G}$. Note that unlike state reversible systems, R-state reversible dynamical systems need not retrace every stage of the original system trajectory in reverse order, nor is it necessary for the dynamical system to recover the initial system state. The function $r(u, y)$ in Definition 2.9 is a generalized *power supply* from the environment to the dynamical system through the system's input-output ports (u, y). Hence, (2.23) assures that the total generalized energy supplied to the dynamical system $\mathcal{G}$

by the environment is returned to the environment over a given R-reversible trajectory starting and ending at any given (not necessarily the same) state $z \in \mathcal{B}$. Furthermore, condition (2.23) assures that a reversible process completely restores the original dynamic state of a system and at the same time restores the energy supplied by the environment back to its original condition. The following result provides sufficient conditions for the existence of an R-reversible trajectory of a dynamical system $\mathcal{G}$, and hence, establishes sufficient conditions for R-state reversibility of the dynamical system $\mathcal{G}$.

Theorem 2.7 *Consider the dynamical system $\mathcal{G}$ defined on $\mathcal{B}$. Let $R : \mathcal{B} \rightarrow \mathcal{B}$ be an involutive operator, and let $s^z(\cdot, u(\cdot)) \in \mathcal{W}_{[t_0, t_1]}$, where $u(\cdot) \in \mathcal{U}$. Assume there exist a continuous function $V : \mathcal{B} \rightarrow \mathbb{R}$ and a function $r : \mathcal{U} \times \mathcal{Y} \rightarrow \mathbb{R}$ such that $V(z) = V(Rz)$, $z \in \mathcal{B}$, and for every $z \in \mathcal{B}$ and all $\hat{t}_0$, $\hat{t}_1$, $t_0 \leq \hat{t}_0 < \hat{t}_1 \leq t_1$,*

$$V(s^z(\hat{t}_1, u(\hat{t}_1))) \geq V(s^z(\hat{t}_0, u(\hat{t}_0))) + \int_{\hat{t}_0}^{\hat{t}_1} r(u(t), y(t)) \mathrm{d}t. \quad (2.24)$$

Furthermore, assume there exists $\mathcal{M} \subset \mathcal{B}$ such that for all $\hat{t}_0$, $\hat{t}_1$, $t_0 \leq \hat{t}_0 < \hat{t}_1 \leq t_1$, and $s^z(t, u(t)) \notin \mathcal{M}$, $t \in [\hat{t}_0, \hat{t}_1]$, (2.24) holds as a strict inequality. If $s^z(\cdot, u(\cdot))$ is an R-reversible trajectory of $\mathcal{G}$, then $s^z(t, u(t)) \in \mathcal{M}$, $t \in [t_0, t_1]$.

Proof. Let $s^z(\cdot, u(\cdot)) \in \mathcal{W}_{[t_0, t_1]}$, where $u(\cdot) \in \mathcal{U}$, be an R-reversible trajectory of $\mathcal{G}$ so that there exists $u^-(\cdot) \in \mathcal{U}$ such that $s^{-z}(\cdot, u^-(\cdot)) \in \mathcal{W}_{[t_0, t_1]}$. Suppose, *ad absurdum*, there exists $t \in [t_0, t_1]$ such that $s^z(t, u(t)) \notin \mathcal{M}$. Now, it follows that there exists an interval $[\hat{t}_0, \hat{t}_1] \subset [t_0, t_1]$ such that for $t_0 \leq \hat{t}_0 < \hat{t}_1 \leq t_1$,

$$V(s^z(\hat{t}_1, u(\hat{t}_1))) > V(s^z(\hat{t}_0, u(\hat{t}_0))) + \int_{\hat{t}_0}^{\hat{t}_1} r(u(t), y(t)) \mathrm{d}t, \quad (2.25)$$

which further implies that

$$V(s^z(t_1, u(t_1))) > V(s^z(t_0, u(t_0))) + \int_{t_0}^{t_1} r(u(t), y(t)) \mathrm{d}t. \quad (2.26)$$

Next, since $s^{-z}(\cdot, u^-(\cdot)) \in \mathcal{W}_{[t_0, t_1]}$, where $u^-(\cdot) \in \mathcal{U}$, it follows that

$$V(s^{-z}(t_1, u^-(t_1))) \geq V(s^{-z}(t_0, u^-(t_0))) + \int_{t_0}^{t_1} r(u^-(t), y^-(t)) \mathrm{d}t.$$

$$(2.27)$$

Now, adding (2.26) and (2.27), using the definition of $s^{-z}(\cdot, u^{-}(\cdot))$, using the fact that $V(z) = V(Rz)$, $z \in \mathcal{B}$, and using (2.23) yields

$$V(s^{z}(t_0, u(t_0))) + V(s^{z}(t_1, u(t_1))) > V(s^{z}(t_0, u(t_0)))$$
$$+V(s^{z}(t_1, u(t_1))), \qquad (2.28)$$

which is a contradiction. Hence, $s^{z}(t, u(t)) \in \mathcal{M}$, $t \in [t_0, t_1]$. $\qquad \square$

It is important to note that since $V : \mathcal{B} \to \mathbb{R}$ in Theorem 2.7 is not sign definite, Theorem 2.7 also holds for the case where the inequality in (2.24) is reversed. The following corollary to Theorem 2.7 is immediate.

Corollary 2.4 *Consider the dynamical system $\mathcal{G}$ defined on $\mathcal{B}$. Let $R : \mathcal{B} \to \mathcal{B}$ be an involutive operator, let $\mathcal{M} \subset \mathcal{B}$, and let $s^{z}(\cdot, u(\cdot)) \in \mathcal{W}_{[t_0, t_1]}$, where $u(\cdot) \in \mathcal{U}$. Assume there exists a continuous function $V : \mathcal{B} \to \mathbb{R}$ such that $V(z) = V(Rz)$, $z \in \mathcal{B}$, and for $s^{z}(t, u(t)) \notin \mathcal{M}$, $t \in [t_1, t_2]$, $V(s(t, z_0, u(\cdot)))$ is a strictly increasing (respectively, decreasing) function of time. If $s^{z}(\cdot, u(\cdot))$ is an R-reversible trajectory of $\mathcal{G}$, then $s^{z}(t, u(t)) \in \mathcal{M}$, $t \in [t_0, t_1]$.*

Proof. The proof is a direct consequence of Theorem 2.7 with $r(u, y) \equiv 0$ and the fact that Theorem 2.7 also holds for the case when the inequality in (2.24) is reversed. $\qquad \square$

It follows from Corollary 2.4 that if, for a given dynamical system $\mathcal{G}$, there exists an R-reversible trajectory of $\mathcal{G}$, then there does not exist a function of the state of the system that strictly decreases or strictly increases in time on any trajectory of $\mathcal{G}$ lying in $\mathcal{M}$. In this case, the existence of a completely ordered time set having a topological structure involving a closed set homeomorphic to the real line cannot be established. Such systems, which include lossless Newtonian and Hamiltonian systems, are time-reversal symmetric and hence lack an inherent time direction. As we see in Sections 3.5 and 5.1, that is not the case with thermodynamic systems.

Next, we present a notion of state recoverability of a dynamical system $\mathcal{G}$.

Definition 2.10 *Consider the dynamical system $\mathcal{G}$ defined on $\mathcal{B}$. Let $r : \mathcal{U} \times \mathcal{Y} \to \mathbb{R}$, and let $s^{z}(\cdot, u(\cdot)) \in \mathcal{W}_{[t_0, t_1]}$, where $u(\cdot) \in \mathcal{U}$. $s^{z}(\cdot, u(\cdot))$ is a recoverable trajectory of $\mathcal{G}$ if there exists $u^{-}(\cdot) \in \mathcal{U}$ and $t_2 > t_1$ such that $u^{-} : [t_1, t_2] \to U$,*

$$s(t_2, s^{z}(t_1, u(t_1)), u^{-}(t_2)) = s^{z}(t_0, u(t_0)), \qquad (2.29)$$

and

$$\int_{t_0}^{t_1} r(u(t), y(t))\mathrm{dt} + \int_{t_1}^{t_2} r(u^-(t), y^-(t))\mathrm{dt} = 0, \qquad (2.30)$$

where $y^-(\cdot)$ denotes the read-out map for the trajectory $s(\cdot, s^z(t_1, u(t_1)), u^-(\cdot))$. Furthermore, $\mathcal{G}$ is a state recoverable dynamical system if for every $z \in \mathcal{B}$, $s^z(\cdot, u(\cdot))$ is a recoverable trajectory of $\mathcal{G}$.

It follows from the definition of state recoverability that the way in which the initial dynamical system state is restored may be chosen freely so long as (2.30) is satisfied. Hence, unlike R-state reversibility, it is not necessary for the dynamical system to recover the initial state of the system through an involutive transformation of the system trajectory. Furthermore, unlike state reversibility, it is not necessary for the dynamical system to retrace every stage of the original trajectory in the reverse order. However, condition (2.30) assures that the recoverable process completely restores the original dynamic state and at the same time restores the energy supplied by the environment back to its original condition. This notion of recoverability is closely related to Planck's notion of complete reversibility, wherein the initial system state is restored in the *totality of Nature* ("die gesamte Natur"). The following result provides a sufficient condition for the existence of a recoverable trajectory of a dynamical system $\mathcal{G}$, and hence, establishes sufficient conditions for state recoverability of $\mathcal{G}$.

Theorem 2.8 *Consider the dynamical system $\mathcal{G}$ defined on $\mathcal{B}$. Let $s^z(\cdot, u(\cdot)) \in \mathcal{W}_{[t_0, t_1]}$, where $u(\cdot) \in \mathcal{U}$. Assume there exist a continuous function $V : \mathcal{B} \to \mathbb{R}$ and a function $r : \mathcal{U} \times \mathcal{Y} \to \mathbb{R}$ such that for every $z \in \mathcal{B}$ and all $\hat{t}_0, \hat{t}_1, t_0 \leq \hat{t}_0 < \hat{t}_1 \leq t_1$,*

$$V(s^z(\hat{t}_1, u(\hat{t}_1))) \geq V(s^z(\hat{t}_0, u(\hat{t}_0))) + \int_{\hat{t}_0}^{\hat{t}_1} r(u(t), y(t))\mathrm{dt}. \quad (2.31)$$

Furthermore, assume there exists $\mathcal{M} \subset \mathcal{B}$ such that for all $\hat{t}_0, \hat{t}_1, t_0 \leq \hat{t}_0 < \hat{t}_1 \leq t_1$, and $s^z(t, u(t)) \notin \mathcal{M}$, $t \in [\hat{t}_0, \hat{t}_1]$, (2.31) holds as a strict inequality. If $s^z(\cdot, u(\cdot))$ is a recoverable trajectory of $\mathcal{G}$, then $s^z(t, u(t)) \in \mathcal{M}$, $t \in [t_0, t_1]$.

Proof. Let $s^z(\cdot, u(\cdot)) \in \mathcal{W}_{[t_0, t_1]}$, where $u(\cdot) \in \mathcal{U}$, be a recoverable trajectory of $\mathcal{G}$ so that there exist $u^-(\cdot) \in \mathcal{U}$ and $t_2 > t_1$ such that $s(t_2, s^z(t_1, u(t_1)), u^-(t_2)) = s^z(t_0, u(t_0))$. Suppose, *ad absurdum*, there exists $t \in [t_0, t_1]$ such that $s^z(t, u(t)) \notin \mathcal{M}$. Now, it follows that there exists an interval $[\hat{t}_0, \hat{t}_1] \subset [t_0, t_1]$ such that for $t_0 \leq \hat{t}_0 < \hat{t}_1 \leq t_1$,

$$V(s^z(\hat{t}_1, u(\hat{t}_1))) > V(s^z(\hat{t}_0, u(\hat{t}_0))) + \int_{\hat{t}_0}^{\hat{t}_1} r(u(t), y(t))dt, \quad (2.32)$$

which further implies that

$$V(s^z(t_1, u(t_1))) > V(s^z(t_0, u(t_0))) + \int_{t_0}^{t_1} r(u(t), y(t))dt. \quad (2.33)$$

Next, it follows from (2.31) with $t_2 > t_1$ that

$$V(s(t_2, s^z(t_1, u(t_1)), u^-(t_2))) \geq V(s(t_1, s^z(t_1, u(t_1)), u^-(t_1)))$$
$$+ \int_{t_1}^{t_2} r(u^-(t), y^-(t))dt. \quad (2.34)$$

Now, adding (2.33) and (2.34), using the definition of $s(t_2, s^z(t_1, u(t_1)), u^-(t_2)))$, and using (2.30) yields

$$V(s^z(t_0, u(t_0))) + V(s^z(t_1, u(t_1))) > V(s^z(t_0, u(t_0)))$$
$$+V(s^z(t_1, u(t_1))), \quad (2.35)$$

which is a contradiction. Hence, $s^z(t, u(t)) \in \mathcal{M}, \ t \in [t_0, t_1]$. □

The following corollary to Theorem 2.8 is immediate.

Corollary 2.5 *Consider the dynamical system $\mathcal{G}$ defined on $\mathcal{B}$. Let $\mathcal{M} \subset \mathcal{B}$, and let $s^z(\cdot, u(\cdot)) \in \mathcal{W}_{[t_0, t_1]}$, where $u(\cdot) \in \mathcal{U}$. Assume there exists a continuous function $V : \mathcal{B} \to \mathbb{R}$ such that for $s^z(t, u(t)) \notin \mathcal{M}, \ t \in [t_0, t_1], \ V(s(t, z_0, u(\cdot))$ is a strictly increasing (respectively, decreasing) function of time. If $s^z(\cdot, u(\cdot))$ is a recoverable trajectory of $\mathcal{G}$, then $s^z(t, u(t)) \in \mathcal{M}, \ t \in [t_0, t_1]$.*

Proof. The proof is a direct consequence of Theorem 2.8 with $r(u, y) \equiv 0$ and the fact that Theorem 2.8 also holds for the case when the inequality in (2.31) is reversed. □

As in the case of R-state reversibility and state reversibility, state recoverability can be used to establish a connection between a dynamical system evolving on a manifold $\mathcal{M} \subset \mathcal{B}$ and the arrow of time. However, in the case of state recoverability, the recoverable dynamical system trajectory need not involve an involutive transformation of the system trajectory, nor is it required to retrace the original system trajectory in recovering the original dynamic state. It should be noted here that state recoverability is not implied by the concepts of *reachability* and *controllability*, which play a central role in control theory (see Section 3.2). For example, one might envision, albeit with a considerable stretch of the imagination, perfectly controlled inputs that could reassemble a broken egg or even fuse water into solid cubes of ice. However, in all such cases, an external source of energy from the environment would be required to operate such an immaculate state recoverable mechanism and would violate condition (2.30). Clearly, state recoverability is a weaker notion than that of state reversibility since state reversibility implies state recoverability; the converse, however, is not generally true. Conversely, state irrecoverability is a logically stronger notion than state irreversibility since state irrecoverability implies state irreversibility. However, as we see in Chapter 3, these notions are equivalent for thermodynamic systems.

2.4 Reversible Dynamical Systems, Volume-Preserving Flows, and Poincaré Recurrence

The notion of R-state reversibility introduced in Section 2.3 is one of the fundamental symmetries that arise in natural science. This notion can also be characterized by the flow of a dynamical system. In particular, consider the dynamical system (2.1) given by

$$\dot{z}(t) = w(z(t)), \quad z(t_0) = z_0, \quad t \in \mathcal{I}_{z_0}, \tag{2.36}$$

where $z(t) \in \mathcal{D} \subseteq \mathbb{R}^q$, $t \in \mathcal{I}_{z_0}$, is the system state vector, $\mathcal{D}$ is an open subset of $\mathbb{R}^q$, and $w : \mathcal{D} \to \mathbb{R}^q$ is locally Lipschitz continuous on $\mathcal{D}$. Since $w(\cdot)$ is locally Lipschitz continuous on $\mathcal{D}$, it follows from Theorem 3.1 of [50, p. 18] that there exists a unique solution to (2.36). In this case, the semi-group property $s(t+\tau, z_0) = s(t, s(\tau, z_0))$, $t+\tau$, $\tau \in \mathcal{I}_{z_0}$ and $t \in \mathcal{I}_{s(\tau,z_0)}$, and the continuity of $s(t, \cdot)$ on $\mathcal{D}$, $t \in \mathcal{I}_{z_0}$, hold. Now, in terms of the flow $s_t : \mathcal{D} \to \mathcal{D}$ of (2.36), the consistency and semi-group properties of (2.36) can be equivalently written as $s_{t_0}(z_0) = z_0$ and $(s_\tau \circ s_t)(z_0) = s_\tau(s_t(z_0)) = s_{t+\tau}(z_0)$, where "$\circ$" denotes the composition operator. Next, it follows from continuity of solutions and the semi-group property that the map $s_t : \mathcal{D} \to \mathcal{D}$ is a continuous function with a continuous inverse s_{-t}. Thus, s_t, $t \in \mathcal{I}_{z_0}$, generates a one-parameter family of homeomorphisms on $\mathcal{D}$ forming a commutative group under composition.

To show that R-state reversibility can be characterized by the flow of (2.36), let $\mathcal{R} : \mathcal{D} \rightarrow \mathcal{D}$ be a continuous map of (2.36) such that

$$\dot{\mathcal{R}}(z(t)) = -w(\mathcal{R}(z(t))), \quad \mathcal{R}(z(t_0)) = \mathcal{R}(z_0), \quad t \in \mathcal{I}_{\mathcal{R}(z_0)}. \quad (2.37)$$

Now, it follows from (2.37) that

$$\mathcal{R} \circ s_t = s_{-t} \circ \mathcal{R}, \quad t \in \mathcal{I}_{z_0}. \quad (2.38)$$

Condition (2.38), with $\mathcal{R}(\cdot)$ satisfying (2.37), defines an R-reversed trajectory of (2.36) in the sense of Definition 2.8 with $\tau(t) = t$.

In the context of classical mechanics involving the *configuration manifold* (space of generalized positions) $\mathcal{Q} = \mathbb{R}^n$, with governing equations given by

$$\dot{q}(t) = \left(\frac{\partial \mathcal{H}(q(t), p(t))}{\partial p(t)} \right)^{\mathrm{T}}, \quad q(t_0) = q_0, \quad t \geq t_0, \quad (2.39)$$

$$\dot{p}(t) = -\left(\frac{\partial \mathcal{H}(q(t), p(t))}{\partial q(t)} \right)^{\mathrm{T}}, \quad p(t_0) = p_0, \quad (2.40)$$

where $q \in \mathbb{R}^n$ denotes generalized system positions, $p \in \mathbb{R}^n$ denotes generalized system momenta, $\mathcal{H} : \mathbb{R}^n \times \mathbb{R}^n \rightarrow \mathbb{R}$ is the system Hamiltonian given by $\mathcal{H}(q, p) \triangleq \dot{q}^{\mathrm{T}} p - \mathcal{L}(q, \dot{q})$, $\mathcal{L}(q, \dot{q})$ is the system Lagrangian,[3] and $p(q, \dot{q}) \triangleq \left(\frac{\partial \mathcal{L}(q, \dot{q})}{\partial \dot{q}} \right)^{\mathrm{T}}$, the reversing symmetry $\mathcal{R} : \mathbb{R}^n \times \mathbb{R}^n \rightarrow \mathbb{R}^n \times \mathbb{R}^n$ is such that $\mathcal{R}(q, p) = (q, -p)$ and satisfies (2.37). In this case, $\mathcal{R}$ is an involution. This implies that if $(q(t), p(t))$, $t \geq t_0$, is a solution to (2.39) and (2.40), then $(q(-t), -p(-t))$, $t \geq t_0$, is also a solution to (2.39) and (2.40) with initial condition $(q_0, -p_0)$. In the configuration space this clearly shows the time reversal nature of lossless mechanical systems.

Reversible dynamical systems tend to exhibit a phenomenon known as *Poincaré recurrence* [4]. Poincaré recurrence states that if a dynamical system has a fixed total energy that restricts its dynamics to bounded subsets of its state space, then the dynamical system will eventually return arbitrarily close to its initial system state infinitely often. More precisely, Poincaré [84] established the fact that if the flow of a dynamical system preserves volume and has only bounded orbits, then for each open set there exist orbits that intersect the set infinitely often. In order to state the Poincaré recurrence theorem, the following definitions are needed.

---

[3]Here we assume that the system Lagrangian is *hyperregular* [64] so that the map from the generalized velocities $\dot{q}$ to the generalized momenta p is *bijective* (i.e., one-to-one and onto).

Definition 2.11 *Let* $\mathcal{V} \subset \mathbb{R}^q$ *be a bounded set. The* volume $\mathcal{V}_{\text{vol}}$ *of* $\mathcal{V}$ *is defined as*

$$\mathcal{V}_{\text{vol}} \triangleq \int_{\mathcal{V}} d\mathcal{V}. \tag{2.41}$$

Definition 2.12 *Let* $\mathcal{V} \subset \mathbb{R}^q$ *be a bounded set. A map* $g : \mathcal{V} \to \mathcal{Q}$, *where* $\mathcal{Q} \subset \mathbb{R}^q$, *is* volume-preserving *if for any* $\mathcal{V}_0 \subset \mathcal{V}$, *the volume of* $g(\mathcal{V}_0)$ *is equal to the volume of* $\mathcal{V}_0$.

The following theorem, known as Liouville's theorem [4], establishes sufficient conditions for volume-preserving flows. For the statement of this theorem, consider the nonlinear dynamical system (2.36) and define the divergence of $w = [w_1, ..., w_q]^{\text{T}} : \mathcal{D} \to \mathbb{R}^q$ by

$$\nabla \cdot w(z) \triangleq \sum_{i=1}^{q} \frac{\partial w_i(z)}{\partial z_i}, \tag{2.42}$$

where ∇ denotes the nabla operator, " $\cdot$ " denotes the dot product in $\mathbb{R}^q$, and z_i denotes the ith element of z.

Theorem 2.9 *Consider the nonlinear dynamical system (2.36). If* $\nabla \cdot w(z) \equiv 0$, *then the flow* $s_t : \mathcal{D} \to \mathcal{D}$ *of (2.36) is volume-preserving.*

Proof. Let $\mathcal{V} \subset \mathbb{R}^q$ be a compact set such that its image at time t under the mapping $s_t(\cdot)$ is given by $s_t(\mathcal{V})$. In addition, let $d\mathcal{S}_{\mathcal{V}}$ denote an infinitesimal surface element of the boundary of the set $\mathcal{V}$ and let $\hat{n}(z)$, $z \in \partial\mathcal{V}$, denote an outward normal vector to the boundary of $\mathcal{V}$. Then the change in volume of $s_t(\mathcal{V})$ at $t = t_0$ is given by

$$ds_t(\mathcal{V})_{\text{vol}} = \int_{\partial\mathcal{V}} (w(z) \cdot \hat{n}(z)) dt d\mathcal{S}_{\mathcal{V}}, \tag{2.43}$$

which, using divergence theorem, implies that

$$\left. \frac{ds_t(\mathcal{V})_{\text{vol}}}{dt} \right|_{t=t_0} = \int_{\partial\mathcal{V}} (w(z) \cdot \hat{n}(z)) d\mathcal{S}_{\mathcal{V}} = \int_{\mathcal{V}} \nabla \cdot w(z) d\mathcal{V}. \tag{2.44}$$

Hence, if $\nabla \cdot w(z) \equiv 0$, then $s_t(\cdot)$ is a volume-preserving map. $\square$

Volume preservation is the key conservation law underlying statistical mechanics. The flows of volume-preserving dynamical systems belong to one of the Lie pseudogroups[4] of diffeomorphisms. These

---

[4]A *Lie group* is a topological group that can be given an analytic structure such that the group operation and inversion are analytic. A *Lie pseudogroup* is an infinite dimensional counterpart of a Lie group.

systems arise in incompressible fluid dynamics, classical mechanics, and acoustics. Next, we state the well known Poincaré recurrence theorem. For this result, let $g^{(n)}(z)$, $n \in \overline{\mathbb{Z}}_+$, denote the n-time composition operator of $g(z)$ with itself and define $g^{(0)}(z) \triangleq z$.

Theorem 2.10 *Let $\mathcal{D} \subset \mathbb{R}^q$ be an open bounded set, and let $g : \mathcal{D} \to \mathcal{D}$ be a continuous, volume-preserving bijective (one-to-one and onto) map. Then for every open set $\mathcal{N} \subset \mathcal{D}$, there exists $n \in \mathbb{Z}_+$ such that $g^{(n)}(\mathcal{N}) \cap \mathcal{N} \neq \emptyset$. Furthermore, there exists a point $z \in \mathcal{N}$ which returns to $\mathcal{N}$, that is, $g^{(n)}(z) \in \mathcal{N}$ for some $n \in \mathbb{Z}_+$.*

Proof. The proof of this result is standard; see for example [4, p. 72]. For completeness of exposition, however, we provide a proof here. First, note that the images $g^{(p)}(\mathcal{N})$, $p \in \overline{\mathbb{Z}}_+$, under the mapping $g(\cdot)$ of the neighborhood $\mathcal{N} \subset \mathcal{D}$ have the same volume and are all contained in $\mathcal{D}$. Next, define the union of all the images of $\mathcal{N}$ by

$$\mathcal{V} \triangleq \bigcup_{p=0}^{\infty} g^{(p)}(\mathcal{N}) \subset \mathcal{D}. \qquad (2.45)$$

Since the volume of a union of disjoint sets is the sum of the individual set volumes, it follows that if $g^{(p)}(\mathcal{N})$, $p \in \overline{\mathbb{Z}}_+$, are disjoint, then $\mathcal{V}_{\text{vol}} = \infty$. However, $\mathcal{V} \subset \mathcal{D}$ and $\mathcal{D}$ is a bounded set by assumption. Hence, there exist $k, l \in \overline{\mathbb{Z}}_+$, with $k > l$, such that $g^{(k)}(\mathcal{N}) \cap g^{(l)}(\mathcal{N}) \neq \emptyset$. Now, applying the inverse $g^{(-1)}$ to this relation l times and using the fact that $g(\cdot)$ is a bijective map, it follows that $g^{(k-l)}(\mathcal{N}) \cap \mathcal{N} \neq \emptyset$. Thus, $g^{(n)}(\mathcal{N}) \cap \mathcal{N} \neq \emptyset$, where $n = k - l$. Hence, there exists a point $z \in \mathcal{N}$ such that $g^{(n)}(z) \in g^{(n)}(\mathcal{N}) \cap \mathcal{N} \subseteq \mathcal{N}$. $\square$

The next result establishes the existence of a point z in $\mathcal{D} \subset \mathbb{R}^q$ such that $\lim_{i \to \infty} g^{(n_i)}(z) = z$ for some sequence $\{n_i\}_{i=1}^{\infty}$, with $n_i \to \infty$ as $i \to \infty$, under a continuous, volume-preserving bijective mapping $g(\cdot)$ which maps a bounded region $\mathcal{D}$ of a Euclidian space onto itself. Hence, z returns infinitely often to any open neighborhood of itself under the mapping $g(\cdot)$.

Theorem 2.11 *Let $\mathcal{D} \subset \mathbb{R}^q$ be an open bounded set, and let $g : \mathcal{D} \to \mathcal{D}$ be a continuous, volume-preserving bijective map. Then for every open neighborhood $\mathcal{N} \subset \mathcal{D}$, there exists a point $z \in \mathcal{N}$ such that $\lim_{i \to \infty} g^{(n_i)}(z) = z$ for some sequence $\{n_i\}_{i=1}^{\infty}$, with $n_i \to \infty$ as $i \to \infty$. Hence, $z \in \mathcal{N}$ returns to $\mathcal{N}$ infinitely often, that is, there exists a sequence $\{n_i\}_{i=1}^{\infty}$, with $n_i \to \infty$ as $i \to \infty$, such that $g^{(n_i)}(z) \in \mathcal{N}$ for all $i \in \mathbb{Z}_+$.*

Proof. Let $\mathcal{N} \subset \mathcal{D}$ be an open set, and let $\mathcal{N}_1 \triangleq \mathcal{B}_{\delta_1}(x_1)$ be such that $\overline{\mathcal{N}}_1 \subset \mathcal{N}$ for some $\delta_1 > 0$ and $x_1 \in \mathcal{N}$. Applying Theorem 2.10, with $g(\cdot)$ replaced by $g^{(-1)}(\cdot)$, it follows that there exists $n_1 \in \mathbb{Z}_+$ such that $g^{(-n_1)}(\mathcal{N}_1) \cap \mathcal{N}_1 \neq \varnothing$, which implies that $g^{(-n_1)}(\mathcal{N}_1) \cap \overline{\mathcal{N}}_1 \neq \varnothing$. Now, let $\mathcal{N}_2 = \mathcal{B}_{\delta_2}(x_2)$ be such that $\overline{\mathcal{N}}_2 \subset g^{(-n_1)}(\mathcal{N}_1) \cap \mathcal{N}_1$ for some $\delta_2 > 0$ and $x_2 \in g^{(-n_1)}(\mathcal{N}_1) \cap \mathcal{N}_1$. Repeating the above arguments it follows that there exists $n_2 \in \mathbb{Z}_+$, $n_2 > n_1$, such that $g^{(-n_2)}(\mathcal{N}_2) \cap \mathcal{N}_2 \neq \varnothing$ and $g^{(-n_2)}(\overline{\mathcal{N}}_2) \cap \overline{\mathcal{N}}_2 \neq \varnothing$. Repeating this process recursively, it follows that there exist sequences $\{n_i\}_{i=1}^{\infty}$ and $\{\delta_i\}_{i=1}^{\infty}$, with $n_i \to \infty$ as $i \to \infty$, $\delta_i \to 0$ as $i \to \infty$, and $\delta_i > \delta_{i+1}$, $i = 1, 2, ...$, such that $\mathcal{N}_i \supset \mathcal{N}_{i+1}$, $i = 1, 2, ...$, and $g^{(-n_i)}(\mathcal{N}_i) \cap \mathcal{N}_i \neq \varnothing$, where $\mathcal{N}_i = \mathcal{B}_{\delta_i}(x_i)$ for some $x_i \in g^{(-n_{i-1})}(\mathcal{N}_{i-1}) \cap \mathcal{N}_{i-1}$ and where $n_0 \triangleq 0$ and $\mathcal{N}_0 \triangleq \mathcal{N}$. Now, since $\mathcal{N}_i \neq \varnothing$, $i \in \mathbb{Z}_+$, it follows from the Cantor intersection theorem [3, p.56] that $\mathcal{Z} \triangleq \bigcap_{i=1}^{\infty} \overline{\mathcal{N}}_i \neq \varnothing$. Furthermore, since $\delta_i \to 0$ as $i \to \infty$, it follows that $\mathcal{Z}$ is a singleton. Next, let $z \in \mathcal{Z} = \{z\}$, and since for every $i \in \mathbb{Z}_+$, $\overline{\mathcal{N}}_{i+1} \subset \mathcal{N}_i$, it follows that $z \in \mathcal{N}_i$, $i \in \mathbb{Z}_+$. Now, note that $z \in \mathcal{N}_{i+1} \subset g^{(-n_i)}(\mathcal{N}_i) \cap \mathcal{N}_i$ for all $i \in \mathbb{Z}_+$, which implies that $g^{(n_i)}(z) \in \mathcal{N}_i$, $i \in \mathbb{Z}_+$. Hence, since $\delta_i \to 0$ as $i \to \infty$, it follows that $\lim_{i \to \infty} g^{(n_i)}(z) = z$. $\square$

The next theorem strengthens Poincaré's theorem by showing that for every open neighborhood $\mathcal{N}$ of $\mathcal{D} \subset \mathbb{R}^q$, there exists a subset of $\mathcal{N}$ that is dense[5] in $\mathcal{N}$ so that almost every moving point in $\mathcal{N}$ returns repeatedly to the vicinity of its initial position under a continuous, volume-preserving bijective mapping which maps the bounded region $\mathcal{D}$ onto itself.

Theorem 2.12 *Let $\mathcal{D} \subset \mathbb{R}^q$ be an open bounded set, and let $g : \mathcal{D} \to \mathcal{D}$ be a continuous, volume-preserving bijective map. Then for every open neighborhood $\mathcal{N} \subset \mathcal{D}$, there exists a dense subset $\mathcal{V} \subset \mathcal{N}$ such that for every point $z \in \mathcal{V}$, $\lim_{i \to \infty} g^{(n_i)}(z) = z$ for some sequence $\{n_i\}_{i=1}^{\infty}$, with $n_i \to \infty$ as $i \to \infty$.*

Proof. Let $\mathcal{N} \subset \mathcal{D}$ be an open neighborhood and define $\mathcal{V} \subset \mathcal{N}$ by

$$\mathcal{V} \triangleq \{z \in \mathcal{N} : \text{there exists a sequence } \{n_i\}_{i=1}^{\infty}, \text{ with } n_i \to \infty$$
$$\text{as } i \to \infty, \text{ such that } \lim_{i \to \infty} g^{(n_i)}(z) = z\}. \qquad (2.46)$$

Now, let $z \in \mathcal{N}$ and let $\{\delta_i\}_{i=1}^{\infty}$ be a strictly decreasing positive sequence with $\delta_i \to 0$ as $i \to \infty$ and $\mathcal{B}_{\delta_1}(z) \subset \mathcal{N}$. It follows from Theorem 2.11 that for every $i \in \mathbb{Z}_+$, there exists $z_i \in \mathcal{B}_{\delta_i}(z)$ such that

---

[5]We say that $\mathcal{V}$ is *dense* in $\mathcal{N}$ if $\mathcal{N}$ is contained in the closure of $\mathcal{V}$.

$\lim_{k\to\infty} g^{(n_k)}(z_i) = z_i$ for some sequence $\{n_k\}_{k=1}^{\infty}$, with $n_k \to \infty$ as $k \to \infty$, which implies that $z_i \in \mathcal{V}$, $i \in \mathbb{Z}_+$. Next, since $\lim_{i\to\infty} z_i = z$, it follows that $z \in \overline{\mathcal{V}}$ which implies that $\mathcal{V} \subseteq \mathcal{N} \subset \overline{\mathcal{V}}$, and hence, $\mathcal{V}$ is a dense subset of $\mathcal{N}$. $\square$

It follows from Theorem 2.12 that almost every point in $\mathcal{D} \subset \mathbb{R}^q$ will return infinitely many times to any open neighborhood of itself under a continuous, volume-preserving bijective mapping which maps a bounded region $\mathcal{D}$ of a Euclidean space onto itself. The following theorem provides several equivalent statements for establishing Poincaré recurrence.

Theorem 2.13 *Let $\mathcal{D} \subset \mathbb{R}^q$ be an open bounded set, and let $g : \mathcal{D} \to \mathcal{D}$ be a continuous, bijective map. Then the following statements are equivalent:*

i) For every open set $\mathcal{N} \subset \mathcal{D}$, there exists $n \in \mathbb{Z}_+$ such that $g^{(n)}(\mathcal{N}) \cap \mathcal{N} \neq \emptyset$.

ii) For every open set $\mathcal{N} \subset \mathcal{D}$, there exists a point $z \in \mathcal{N}$ which returns to $\mathcal{N}$, that is, $g^{(n)}(z) \in \mathcal{N}$ for some $n \in \mathbb{Z}_+$.

iii) For every open set $\mathcal{N} \subset \mathcal{D}$, there exists a point $z \in \mathcal{N}$ which returns to $\mathcal{N}$ infinitely often, that is, $g^{(n_i)}(z) \in \mathcal{N}$, $i \in \mathbb{Z}_+$, for some sequence $\{n_i\}_{i=1}^{\infty}$, with $n_i \to \infty$ as $i \to \infty$.

iv) For every open set $\mathcal{N} \subset \mathcal{D}$, there exists a point $z \in \mathcal{N}$ such that $\lim_{i\to\infty} g^{(n_i)}(z) = z$ for some sequence $\{n_i\}_{i=1}^{\infty}$, with $n_i \to \infty$ as $i \to \infty$.

v) For every open set $\mathcal{N} \subset \mathcal{D}$, there exists a dense subset $\mathcal{V} \subset \mathcal{N}$ such that for every point $z \in \mathcal{V}$, $\lim_{i\to\infty} g^{(n_i)}(z) = z$ for some sequence $\{n_i\}_{i=1}^{\infty}$, with $n_i \to \infty$ as $i \to \infty$.

Proof. The equivalence of *i)* and *ii)* as well as the implications *iii)* implies *ii)*, *iv)* implies *iii)*, and *v)* implies *iv)* follow trivially. The proof of *i)* implies *iv)* is identical to that of Theorem 2.11, and the proof of *iv)* implies *v)* is identical to that of Theorem 2.12. $\square$

Note that it follows from Theorems 2.10, 2.11, and 2.12 that a continuous, bijective map $g : \mathcal{D} \to \mathcal{D}$ exhibits Poincaré recurrence (that is, one of the statements in Theorem 2.13 holds) if $g(\cdot)$ is volume-preserving. For the remainder of this section we consider the nonlinear dynamical system (2.36) and assume that the solutions to (2.36) are defined for all $t \in \mathbb{R}$. Recall that if all solutions to (2.36) are bounded, then it follows from the Peano-Cauchy theorem [50, pp. 16, 17] that $\mathcal{I}_{z_0} = \mathbb{R}$. The following theorem shows that if a dynamical system preserves volume, then almost all trajectories return arbitrarily close to their initial position infinitely often.

Theorem 2.14 *Consider the nonlinear dynamical system (2.36). Assume that the flow $s_t : \mathcal{D} \to \mathcal{D}$ of (2.36) is volume-preserving and maps an open bounded set $\mathcal{D}_c \subset \mathbb{R}^q$ onto itself, that is, $\mathcal{D}_c$ is an invariant set with respect to (2.36). Then the nonlinear dynamical system (2.36) exhibits Poincaré recurrence, that is, almost every point $z \in \mathcal{D}_c$ returns to every open neighborhood $\mathcal{N} \subset \mathcal{D}_c$ of z infinitely many times.*

Proof. Since $w : \mathcal{D} \to \mathbb{R}^q$ is locally Lipschitz continuous on $\mathcal{D}$ and $s_t(\cdot)$ maps an open bounded set $\mathcal{D}_c \subset \mathbb{R}^n$ onto itself, it follows that the solutions to (2.36) are bounded and unique for all $t \in \mathbb{R}$ and $z_0 \in \mathcal{D}_c$. Thus, the mapping $s_t(\cdot)$ is bijective. Furthermore, since the solutions of (2.36) are continuously dependent on the system's initial conditions, it follows that $s_t(\cdot)$ is continuous. Now, the result follows as a direct consequence of Theorem 2.12 with $g(\cdot) = s_t(\cdot)$ for any $t \geq t_0$. $\square$

It follows from Theorem 2.14 that a nonlinear dynamical system exhibits Poincaré recurrence if one of the statements in Theorem 2.13 holds with $g(\cdot) = s_t(\cdot)$ for any $t \geq t_0$. Note that in this case it follows from *iv*) of Theorem 2.13 that Poincaré recurrence is equivalent to the existence of a point $z \in \mathcal{D}_c$ such that z belongs to its positive limit set, that is, $z \in \omega(z)$.

All Hamiltonian dynamical systems of the form (2.39) and (2.40) exhibit Poincaré recurrence since they possess volume-preserving flows and are conservative in the sense that the Hamiltonian function $\mathcal{H}(q, p)$ remains constant along system trajectories. To see this, note that with $z \triangleq [q^{\mathrm{T}}, p^{\mathrm{T}}]^{\mathrm{T}}$, (2.39) and (2.40) can be rewritten as

$$\dot{z}(t) = \mathcal{J}\left(\frac{\partial \mathcal{H}}{\partial z}(z(t))\right)^{\mathrm{T}}, \quad z(t_0) = z_0, \quad t \geq t_0, \qquad (2.47)$$

where $z_0 \triangleq [q_0^{\mathrm{T}}, p_0^{\mathrm{T}}]^{\mathrm{T}} \in \mathbb{R}^{2n}$ and

$$\mathcal{J} \triangleq \begin{bmatrix} 0_n & I_n \\ -I_n & 0_n \end{bmatrix}. \qquad (2.48)$$

Now, since

$$\dot{\mathcal{H}}(z) = \left(\frac{\partial \mathcal{H}}{\partial z}(z)\right) \mathcal{J} \left(\frac{\partial \mathcal{H}}{\partial z}(z)\right)^{\mathrm{T}} = 0, \quad z \in \mathbb{R}^{2n}, \qquad (2.49)$$

the Hamiltonian function $\mathcal{H}(\cdot)$ is conserved along the flow of (2.47). If $\mathcal{H}(\cdot)$ is bounded from below and is radially unbounded, then every trajectory of the Hamiltonian system (2.47) is bounded. Hence, by choosing the bounded region $\mathcal{D} \triangleq \{z \in \mathbb{R}^{2n} : \mathcal{H}(z) \leq \eta\}$, where $\eta \in \mathbb{R}$ and $\eta > 0$, it follows that the flow $s_t(\cdot)$ of (2.47) maps the bounded region $\mathcal{D}$ onto itself. Since $\eta > 0$ is arbitrary, the region $\mathcal{D}$ can be chosen arbitrarily large. Furthermore, since (2.47) possesses unique solutions over $\mathbb{R}$, it follows that the mapping $s_t(\cdot)$ is one-to-one and onto. Moreover,

$$\nabla \cdot \mathcal{J} \left(\frac{\partial \mathcal{H}}{\partial z}(z)\right)^{\mathrm{T}} = \sum_{i=1}^{n} \frac{\partial^2 \mathcal{H}(q, p)}{\partial q_i \partial p_i} - \sum_{i=1}^{n} \frac{\partial^2 \mathcal{H}(q, p)}{\partial p_i \partial q_i} = 0, \quad z \in \mathbb{R}^{2n},$$

$$(2.50)$$

which, by Theorem 2.9, shows that the flow $s_t(\cdot)$ of (2.47) is volume-preserving. Finally, since the flow $s_t(\cdot)$ of (2.47) is volume-preserving, continuous, and bijective, and $s_t(\cdot)$ maps a bounded region of a Euclidean space onto itself, it follows from Theorem 2.14 that the Hamiltonian dynamical system (2.47) exhibits Poincaré recurrence. That is, in any open neighborhood $\mathcal{N}$ of any point $z_0 \in \mathbb{R}^{2n}$ there exists a point $y \in \mathcal{N}$ such that the trajectory $s(t, y)$, $t \geq t_0$, of (2.47) will return to $\mathcal{N}$ infinitely many times.

Poincaré recurrence has been the main source for the long and fierce debate between the microscopic and macroscopic points of view of thermodynamics. In thermodynamic models predicated on statistical mechanics, an isolated dynamical system will return arbitrarily close to its initial state of molecular positions and velocities infinitely often. If the system entropy is determined by the state variables, then it must also return arbitrarily close to its original value, and hence, undergo cyclical changes. This apparent contradiction between the behavior of a mechanical system of particles and the second law of thermodynamics remains one of the hardest and most controversial problems in statistical physics. The resolution of this paradox lies in the controversial statement that as system dimensionality increases, the recurrence time increases at an extremely fast rate. Nevertheless, the shortcoming of the mechanistic world view of thermodynamics is the absence of the emergence of damping in lossless mechanical systems. The emergence of damping is, however, ubiquitous in isolated thermodynamic systems. Hence, the development of a viable dynamical system model for thermodynamics must guarantee the absence of Poincaré recurrence. The next set of results presents sufficient conditions for the absence of Poincaré recurrence for the nonlinear dynamical system (2.36). For these results define the set of equilibria for the nonlinear dynamical system (2.36) in $\mathcal{D}$ by $\mathcal{M}_e \triangleq \{z \in \mathcal{D} : w(z) = 0\}$.

Theorem 2.15 *Consider the nonlinear dynamical system (2.36) and assume that $\mathcal{D} \setminus \mathcal{M}_e \neq \emptyset$. Assume that there exists a continuous function $V : \mathcal{D} \to \mathbb{R}$ such that for every $z_0 \in \mathcal{D} \setminus \mathcal{M}_e$, $V(s(t, z_0))$, $t \geq t_0$, is a strictly increasing (respectively, decreasing) function of time. Then the nonlinear dynamical system (2.36) does not exhibit Poincaré recurrence in $\mathcal{D} \setminus \mathcal{M}_e$. That is, for some $z \in \mathcal{D} \setminus \mathcal{M}_e$, there exists an open neighborhood $\mathcal{N} \subset \mathcal{D} \setminus \mathcal{M}_e$ of z such that for every $y \in \mathcal{N}$, $y \notin \omega(y)$.*

Proof. Suppose, *ad absurdum*, there exists $z \in \mathcal{D} \setminus \mathcal{M}_e$ such that for every open neighborhood $\mathcal{N}$ containing z, there exists a point $y \in \mathcal{N}$ such that $y \in \omega(y)$. Now, let $\{t_i\}_{i=1}^{\infty}$ be such that $t_i \to \infty$ as $i \to \infty$ and $s(t_i, y) \to y$ as $i \to \infty$. Since $V(\cdot)$ is continuous, it follows that $\lim_{i \to \infty} V(s(t_i, y)) = V(y)$. However, since $V(s(\cdot, y))$ is strictly increasing, it follows that $\lim_{i \to \infty} V(s(t_i, y)) > V(y)$, which is a contradiction. The proof for the case where $V(s(t, x_0))$, $t \geq t_0$, is strictly decreasing is identical. $\square$

For the remainder of this section let $\mathcal{D}_c \subseteq \mathcal{D}$ be a closed invariant set with respect to the nonlinear dynamical system (2.36). The following definitions for convergence and stability with respect to a positively invariant set are needed.

Definition 2.13 *The nonlinear dynamical system (2.36) is convergent with respect to $\mathcal{D}_c$ if $\lim_{t \to \infty} s(t, z)$ exists for every $z \in \mathcal{D}_c$.*

Definition 2.14 ([12]) *An equilibrium point $z \in \mathcal{D}_c \subseteq \mathcal{D}$ of the nonlinear dynamical system (2.36) is Lyapunov stable with respect to the positively invariant set $\mathcal{D}_c$ if, for every relatively open subset $\mathcal{N}_\varepsilon$ of $\mathcal{D}_c$ containing z, there exists a relatively open subset $\mathcal{N}_\delta$ of $\mathcal{D}_c$ containing z such that $s_t(\mathcal{N}_\delta) \subset \mathcal{N}_\varepsilon$ for all $t \geq t_0$. An equilibrium point $z \in \mathcal{D}_c$ of the nonlinear dynamical system (2.36) is semistable if it is Lyapunov stable and there exists a relatively open subset $\mathcal{N}$ of $\mathcal{D}_c$ containing z such that for all initial conditions in $\mathcal{N}$, the trajectory of (2.36) converges to a Lyapunov stable equilibrium point, that is, $\|s(t, x) - y\| \to 0$ as $t \to \infty$, where $y \in \mathcal{D}_c$ is a Lyapunov stable equilibrium point of (2.36) and $x \in \mathcal{N}$. The nonlinear dynamical system*

(2.36) is said to be semistable *if every equilibrium point of (2.36) is semistable.*

If the system (2.36) is convergent with respect to $\mathcal{D}_c$, then the ω-limit set $\omega(z)$ of (2.36) for the trajectory $s^z(t)$ starting at $z \in \mathcal{D}_c$ is a singleton. Furthermore, it follows from continuity of solutions that for every $h \geq 0$, $s_h(\omega(z)) \triangleq \lim_{t \to \infty} s(t + h, z) = \omega(z)$. Thus, $\frac{ds_h(\omega(z))}{dh}\Big|_{h=0} = 0$ and hence $\omega(z)$ is an equilibrium point of (2.36) for all $z \in \mathcal{D}_c$. The next result relates the continuity of the function $\omega(\cdot)$ at a point z to the stability of the equilibrium point $\omega(z)$.

Proposition 2.3 *Suppose the nonlinear dynamical system (2.36) is convergent with respect to $\mathcal{D}_c$. If $\omega(z)$ is a Lyapunov stable equilibrium point for some $z \in \mathcal{D}_c$, then $\omega : \mathcal{D}_c \to \mathcal{D}_c$ is continuous at z.*

Proof. The proof of the result appears in [12]. For completeness of exposition, we provide a proof here. Suppose $\omega(z)$ is Lyapunov stable for some $z \in \mathcal{D}_c$, and let $\mathcal{N}_\varepsilon$ be an open neighborhood of $\omega(z)$. Moreover, choose open neighborhoods $\mathcal{N}$ and $\mathcal{N}_\delta$ of $\omega(z)$ such that $\overline{\mathcal{N}} \subset \mathcal{N}_\varepsilon$ and $s_t(\mathcal{N}_\delta) \subseteq \mathcal{N}$ for all $t \geq t_0$, and let $\{z_i\}_{n=1}^\infty$ be a sequence in $\mathcal{D}_c$ converging to z. The existence of such neighborhoods follows from the Lyapunov stability of $\omega(z)$. Next, there exists $h > t_0$ such that $s(h, z) \in \mathcal{N}_\delta$ and, since the solutions to (2.36) are continuously dependent on the system initial conditions, it follows that there exists an open neighborhood $\mathcal{N}_{\hat{\delta}} \triangleq \mathcal{B}_{\hat{\delta}}(z)$, $\hat{\delta} > 0$, of z such that $s(h, y) \in \mathcal{N}_\delta$ for all $y \in \mathcal{N}_{\hat{\delta}}$. Furthermore, it follows from the Lyapunov stability of $\omega(z)$ that $s(t + h, y) \in \mathcal{N}$, $y \in \mathcal{N}_{\hat{\delta}}$, $t \geq 0$, and hence, $\omega(y) \in \overline{\mathcal{N}} \subset \mathcal{N}_\varepsilon$, $y \in \mathcal{N}_{\hat{\delta}}$, which proves that $\omega : \mathcal{D}_c \to \mathcal{D}_c$ is continuous at z. $\square$

The next result gives an alternative sufficient condition for the absence of Poincaré recurrence in a dynamical system.

Theorem 2.16 *Consider the nonlinear dynamical system (2.36). Assume that $\mathcal{D}_c \setminus \mathcal{M}_e \neq \emptyset$ and assume (2.36) is convergent and semistable in $\mathcal{D}_c$. Then the nonlinear dynamical system (2.36) does not exhibit Poincaré recurrence in $\mathcal{D}_c \setminus \mathcal{M}_e$. That is, for some $z \in \mathcal{D}_c \setminus \mathcal{M}_e$, there exists an open neighborhood $\mathcal{N} \subset \mathcal{D}_c \setminus \mathcal{M}_e$ of z such that for any $y \in \mathcal{N}$ the trajectory $s(t, y)$, $t \geq t_0$, does not return to $\mathcal{N}$ infinitely many times.*

Proof. Let $z \in \mathcal{D}_c \setminus \mathcal{M}_e$ and let $\omega(z) \in \mathcal{M}_e$ be a limiting point for the trajectory $s(t, z)$, $t \geq t_0$, so that $\lim_{t \to \infty} s(t, z) = \omega(z)$. Since (2.36) is convergent and semistable, it follows from Proposition 2.3

that $\omega(z)$, $z \in \mathcal{D}_c \setminus \mathcal{M}_e$, is continuous. Hence, for any $\varepsilon > 0$ there exists $\delta = \delta(\varepsilon) > 0$ such that $\omega(y) \in \mathcal{B}_\varepsilon(\omega(z))$ for all $y \in \mathcal{B}_\delta(z)$. Choose $\varepsilon > 0$ and $\delta > 0$ such that $\overline{\mathcal{B}}_\delta(z) \cap \overline{\mathcal{B}}_\varepsilon(\omega(z)) = \emptyset$. Furthermore, choose $\hat{\varepsilon} > 0$ to be sufficiently small such that

$$\overline{\bigcup_{y \in \mathcal{B}_\delta(z)} \mathcal{B}_{\hat{\varepsilon}}(\omega(y))} \cap \overline{\mathcal{B}}_\delta(z) = \emptyset. \tag{2.51}$$

Since the dynamical system (2.36) is convergent in $\mathcal{D}_c$, it follows that for all $y \in \mathcal{B}_\delta(z)$ and $\hat{\varepsilon} > 0$, there exists $T(\hat{\varepsilon}, y) > t_0$ such that $s(t, y) \in \mathcal{B}_{\hat{\varepsilon}}(\omega(y))$ for all $t > T(\hat{\varepsilon}, y)$. Moreover, it follows from (2.51) that, for all $y \in \mathcal{B}_\delta(z)$, $s(t, y)$, $t \geq t_0$, does not return to $\mathcal{B}_\delta(z)$ infinitely many times, which proves the result with $\mathcal{N} = \mathcal{B}_\delta(z)$. $\square$

Finally, we close this chapter by noting that the results of this section also apply (with minor modifications) to infinite-dimensional dynamical systems $\mathcal{G}$ with flows $s_t : \mathcal{B} \to \mathcal{B}$ defined on a Banach space $\mathcal{B}$. In particular, we require that the map $g(\cdot)$ in the theorems just presented be a measurable transformation $g : \mathcal{B} \to \mathcal{B}$ that preserves a finite measure $\mu(\cdot)$ on $\mathcal{B}$, that is, $\mu(g^{(-1)}(\mathcal{N})) = \mu(\mathcal{N})$ for every measurable set $\mathcal{N} \subset \mathcal{B}$. In this case, the results in this section also hold for infinite-dimensional dynamical systems with this appropriate minor modification.

Chapter Three

A Systems Foundation for Thermodynamics

3.1 Introduction

The fundamental and unifying concept in the analysis of complex (large-scale) dynamical systems is the concept of energy. As noted in Chapter 1, the energy of a state of a dynamical system is the measure of its ability to produce changes (motion) in its own system state as well as changes in the system states of its surroundings. These changes occur as a direct consequence of the energy flow between different subsystems within the dynamical system. Since heat (energy) is a fundamental concept of thermodynamics involving the capacity of hot bodies (more energetic subsystems) to produce work, thermodynamics is a theory of large-scale dynamical systems. As in thermodynamic systems, dynamical systems can exhibit energy (due to friction) that becomes unavailable to do useful work. This in turn contributes to an increase in system entropy, a measure of the tendency of a system to lose the ability to do useful work. In this chapter we use a large-scale dynamical systems perspective to provide a system-theoretic foundation for thermodynamics that bridges the gap between classical and statistical thermodynamics.

To develop a systems foundation for thermodynamics, we use the state space formalism to construct a mathematical model that is consistent with basic thermodynamic principles. This is in sharp contrast to classical thermodynamics wherein an input-output description of the system is used. However, it is felt that a state space formulation is essential for developing a thermodynamic model with enough detail for describing the thermal behavior of heat and dynamical energy. In addition, such a model is crucial in accounting for internal system properties such as compartmental system dynamics characterizing conservation laws wherein subsystem energies can only be transported, stored, or dissipated but not created. If a physical system possesses these conservation properties externally, then there exists a high possibility that the system possesses these properties internally. In this case, assuming that the system possesses these conservation properties internally does not violate any input-output behavior of the physical system established by experimental evidence, and hence, the state space model is theoretically credible. The absence of a state space formalism in classical thermodynamics, and physics in general, is quite disturbing and in our view largely responsible for the monomeric state of classical thermodynamics.

3.2 Conservation of Energy and the First Law of Thermodynamics

To develop a system-theoretic foundation for thermodynamics, we consider a large-scale system model with a combination of subsystems (compartments or parts) that is perceived as a single entity. For each subsystem (compartment) making up the system, we postulate the existence of an *energy* state variable such that the knowledge of these subsystem state variables at any given time $t = t_0$, together with the knowledge of any inputs (heat fluxes) to each of the subsystems for time $t \geq t_0$, completely determines the behavior of the system for any given time $t \geq t_0$. Hence, the (energy) state of our dynamical system at time t is uniquely determined by the state at time t_0 and any external inputs for time $t \geq t_0$ and is independent of the state and inputs before time t_0.

More precisely, we consider a large-scale dynamical system composed of a large number of units with aggregated (or lumped) energy variables representing homogeneous groups of these units. If all the units comprising the system are identical (that is, the system is perfectly homogeneous), then the behavior of the dynamical system can be captured by that of a single plenipotentiary unit. Alternatively, if every interacting system unit is distinct, then the resulting model constitutes a microscopic system. To develop a middle-ground thermodynamic model placed between complete aggregation (classical thermodynamics) and complete disaggregation (statistical thermodynamics), we subdivide the large-scale dynamical system into a finite number of compartments, each formed by a large number of homogeneous units. Each compartment represents the energy content of the different parts of the dynamical system, and different compartments interact by exchanging heat. Thus, our compartmental thermodynamic model utilizes subsystems or compartments to describe the energy distribution among distinct regions in space with intercompartmental flows representing the heat transfer between these regions. Decreasing the number of compartments results in a more aggregated or homogeneous model, whereas increasing the number of compartments leads to a higher degree of disaggregation resulting in a heterogeneous model.

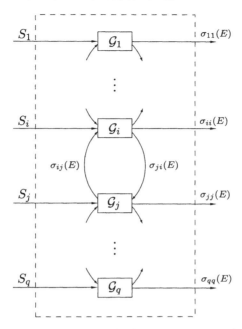

Figure 3.1 Large-scale dynamical system $\mathcal{G}$.

To formulate our state space thermodynamic model, consider the large-scale dynamical system $\mathcal{G}$ shown in Figure 3.1 involving energy exchange between q interconnected subsystems. Let $E_i : [0, \infty) \to \overline{\mathbb{R}}_+$ denote the energy (and hence a nonnegative quantity) of the ith subsystem, let $S_i : [0, \infty) \to \mathbb{R}$ denote the external power (heat flux) supplied to (or extracted from) the ith subsystem, let $\sigma_{ij} : \overline{\mathbb{R}}_+^q \to \overline{\mathbb{R}}_+$, $i \neq j$, $i, j = 1, ..., q$, denote the instantaneous rate of energy (heat) flow from the jth subsystem to the ith subsystem, and let $\sigma_{ii} : \overline{\mathbb{R}}_+^q \to \overline{\mathbb{R}}_+$, $i = 1, ..., q$, denote the instantaneous rate of energy (heat) dissipation from the ith subsystem to the environment. Here we assume that $\sigma_{ij} : \overline{\mathbb{R}}_+^q \to \overline{\mathbb{R}}_+$, $i, j = 1, ..., q$, are locally Lipschitz continuous on $\overline{\mathbb{R}}_+^q$ and $S_i : [0, \infty) \to \mathbb{R}$, $i = 1, ..., q$, are bounded piecewise continuous functions of time.

An *energy balance* for the ith subsystem yields

$$E_i(T) = E_i(t_0) + \sum_{j=1, j \neq i}^{q} \int_{t_0}^{T} [\sigma_{ij}(E(t)) - \sigma_{ji}(E(t))] \mathrm{d}t$$

$$- \int_{t_0}^{T} \sigma_{ii}(E(t)) \mathrm{d}t + \int_{t_0}^{T} S_i(t) \mathrm{d}t, \quad T \geq t_0, \qquad (3.1)$$

or, equivalently, in vector form,

$$E(T) = E(t_0) + \int_{t_0}^{T} w(E(t))\mathrm{d}t - \int_{t_0}^{T} d(E(t))\mathrm{d}t + \int_{t_0}^{T} S(t)\mathrm{d}t, \ T \geq t_0,$$

$$(3.2)$$

where $E(t) \triangleq [E_1(t), ..., E_q(t)]^{\mathrm{T}}$, $d(E(t)) \triangleq [\sigma_{11}(E(t)), ..., \sigma_{qq}(E(t))]^{\mathrm{T}}$, $S(t) \triangleq [S_1(t), ..., S_q(t)]^{\mathrm{T}}$, $t \geq t_0$, and $w = [w_1, ..., w_q]^{\mathrm{T}} : \overline{\mathbb{R}}_+^q \to \mathbb{R}^q$ is such that

$$w_i(E) = \sum_{j=1, j \neq i}^{q} [\sigma_{ij}(E) - \sigma_{ji}(E)], \quad E \in \overline{\mathbb{R}}_+^q. \qquad (3.3)$$

It is important to note that the exchange of energy between subsystems in (3.1) is assumed to be a nonlinear function of all the subsystems, that is, $\sigma_{ij} = \sigma_{ij}(E)$, $E \in \overline{\mathbb{R}}_+^q$, $i \neq j$, $i, j = 1, ..., q$. This assumption is made for generality and would depend on the complexity of the diffusion process. For example, thermal processes may include evaporative and radiative heat transfer as well as thermal conduction giving rise to complex heat transport mechanisms. However, for simple diffusion processes it suffices to assume that $\sigma_{ij}(E) = \sigma_{ij}(E_j)$, wherein the energy flow from the jth subsystem to the ith subsystem is only dependent (possibly nonlinearly) on the energy in the jth subsystem. Similar comments apply to system dissipation.

Note that (3.1) yields a conservation of energy equation and implies that the energy stored in the ith subsystem is equal to the external energy supplied to (or extracted from) the ith subsystem plus the energy gained by the ith subsystem from all other subsystems due to subsystem coupling minus the energy dissipated from the ith subsystem to the environment. Equivalently, (3.1) can be rewritten as

$$\dot{E}_i(t) = \sum_{j=1, j \neq i}^{q} [\sigma_{ij}(E(t)) - \sigma_{ji}(E(t))] - \sigma_{ii}(E(t)) + S_i(t),$$

$$E_i(t_0) = E_{i0}, \quad t \geq t_0, \quad (3.4)$$

or, in vector form,

$$\dot{E}(t) = w(E(t)) - d(E(t)) + S(t), \quad E(t_0) = E_0, \quad t \geq t_0, \quad (3.5)$$

where $E_0 \triangleq [E_{10}, ..., E_{q0}]^{\mathrm{T}}$, yielding a *power balance* equation that characterizes energy flow between subsystems of the large-scale dynamical system $\mathcal{G}$. Equation (3.4) shows that the rate of change of energy, or power, in the ith subsystem is equal to the power input (heat flux) to the ith subsystem plus the energy (heat) flow to the ith subsystem from all other subsystems minus the power dissipated from the ith subsystem to the environment. Furthermore, since $w(\cdot) - d(\cdot)$ is locally Lipschitz continuous on $\overline{\mathbb{R}}_+^q$ and $S(\cdot)$ is a bounded piecewise continuous function of time, it follows that (3.5) has a unique solution over the finite time interval $[t_0, \tau_{E_0})$. If, in addition, the power balance equation (3.5) is *input-to-state stable* [56], then $\tau_{E_0} = \infty$.

Equation (3.2) or, equivalently, (3.5) is a statement of the *first law of thermodynamics* as applied to *isochoric transformations* (i.e., constant subsystem volume transformations) for each of the subsystems $\mathcal{G}_i$, $i = 1, ..., q$, with $E_i(\cdot)$, $S_i(\cdot)$, $\sigma_{ij}(\cdot)$, $i \neq j$, and $\sigma_{ii}(\cdot)$, $i, j = 1, ..., q$, playing the role of the ith subsystem internal energy, rate of heat supplied to (or extracted from) the ith subsystem, heat flow between subsystems due to coupling, and the rate of energy (heat) dissipated to the environment, respectively. To further elucidate that (3.2) is essentially the statement of the principle of the conservation of energy, let the total energy in the large-scale dynamical system $\mathcal{G}$ be given by $U \triangleq \mathbf{e}^{\mathrm{T}} E$, where $\mathbf{e}^{\mathrm{T}} \triangleq [1, ..., 1]$ and $E \in \overline{\mathbb{R}}_+^q$, and let the net energy received by the large-scale dynamical system $\mathcal{G}$ over the time interval $[t_1, t_2]$ be given by

$$Q \triangleq \int_{t_1}^{t_2} \mathbf{e}^{\mathrm{T}} [S(t) - d(E(t))] \mathrm{d}t, \tag{3.6}$$

where $E(t)$, $t \geq t_0$, is the solution to (3.5). Then, premultiplying (3.2) by $\mathbf{e}^{\mathrm{T}}$ and using the fact that $\mathbf{e}^{\mathrm{T}} w(E) \equiv 0$, it follows that

$$\Delta U = Q, \tag{3.7}$$

where $\Delta U \triangleq U(t_2) - U(t_1)$ denotes the variation in the total energy of the large-scale dynamical system $\mathcal{G}$ over the time interval $[t_1, t_2]$. This is a statement of the first law of thermodynamics for isochoric transformations of the large-scale dynamical system $\mathcal{G}$ and gives a precise formulation of the equivalence between the variation in system internal energy and heat.

It is important to note that the large-scale dynamical system model (3.5) does not consider work done by the system on the environment nor work done by the environment on the system. Hence, Q can be physically interpreted as the net amount of energy that is received by the system in forms other than work. The extension of addressing work performed by and on the system can be easily addressed by including an additional state equation, coupled to the power balance equation (3.5), involving volume (deformation) states for each subsystem. Since this extension does not alter any of the conceptual results of this chapter, it is not considered in this chapter for simplicity of exposition. Work performed by the dynamical system on the environment and work done by the environment on the dynamical system is addressed in Chapter 5.

In its most general form thermodynamics can also involve reacting mixtures and combustion. When a chemical reaction occurs, the bonds within molecules of the *reactant* are broken, and atoms and electrons rearrange to form *products*. The thermodynamic analysis of reactive systems can be addressed as an extension of the compartmental thermodynamic model described above. Specifically, in this case the compartments would qualitatively represent different quantities in the same space, and the intercompartmental flows would represent transformation rates in addition to transfer rates. In particular, the compartments would additionally represent quantities of different chemical substances contained within the compartment, and the compartmental flows would additionally characterize transformation rates of reactants into products. The fundamental concepts, however, for developing a thermodynamically consistent state space model would remain the same with an additional mass balance equation included for addressing conservation of energy as well as conservation of mass. This additional mass conservation equation would involve the law of mass action enforcing proportionality between a particular reaction rate and the concentrations of the reactants, and the law of superposition of elementary reactions assuring that the resultant rates for a particular species is the sum of the elementary reaction rates for the species. Even though this is an important extension for developing a general theory for thermodynamics, it does not alter the fundamental conceptual mathematical results developed in this and subsequent chapters and hence is not addressed in this monograph.

For our large-scale dynamical system model $\mathcal{G}$, we assume that $\sigma_{ij}(E) = 0$, $E \in \overline{\mathbb{R}}_+^q$, whenever $E_j = 0$, $i, j = 1, ..., q$. In this case, $w(E) - d(E)$, $E \in \overline{\mathbb{R}}_+^q$, is essentially nonnegative. The above constraint implies that if the energy of the jth subsystem of $\mathcal{G}$ is zero, then this subsystem cannot supply any energy to its surroundings nor dissipate energy to the environment. Moreover, we assume that $S_i(t) \geq 0$ whenever $E_i(t) = 0$, $t \geq t_0$, $i = 1, ..., q$, which implies that when the energy of the ith subsystem is zero, then no energy can be extracted from this subsystem. The following proposition is needed for the main results of this chapter.

Proposition 3.1 *Consider the large-scale dynamical system $\mathcal{G}$ with power balance equation given by (3.5). Suppose $\sigma_{ij}(E) = 0$, $E \in \overline{\mathbb{R}}_+^q$,* *whenever* $E_j = 0$, $i, j = 1, ..., q$, *and* $S_i(t) \geq 0$ *whenever* $E_i(t) = 0$, $t \geq t_0$, $i = 1, ..., q$. *Then the solution* $E(t)$, $t \geq t_0$, *to (3.5) is nonnegative for all nonnegative initial conditions* $E_0 \in \overline{\mathbb{R}}_+^q$.

Proof. First note that $w(E) - d(E)$, $E \in \overline{\mathbb{R}}_+^q$, is essentially nonnegative. Next, since $S_i(t) \geq 0$ whenever $E_i(t) = 0$, $t \geq t_0$, $i = 1, ..., q$, it follows that $\dot{E}_i(t) \geq 0$ for all $t \geq t_0$ and $i = 1, ..., q$ whenever $E_i(t) = 0$ and $E_j(t) \geq 0$ for all $j \neq i$ and $t \geq t_0$. This implies that for all nonnegative initial conditions $E_0 \in \overline{\mathbb{R}}_+^q$, the trajectory of $\mathcal{G}$ is directed towards the interior of the nonnegative orthant $\overline{\mathbb{R}}_+^q$ whenever $E_i(t) = 0$, $i = 1, ..., q$, and hence remains nonnegative for all $t \geq t_0$. $\square$

Next, premultiplying (3.2) by $\mathbf{e}^T$, using Proposition 3.1, and using the fact that $\mathbf{e}^T w(E) \equiv 0$, it follows that

$$\mathbf{e}^T E(T) = \mathbf{e}^T E(t_0) + \int_{t_0}^{T} \mathbf{e}^T S(t) \mathrm{d}t - \int_{t_0}^{T} \mathbf{e}^T d(E(t)) \mathrm{d}t, \quad T \geq t_0.$$
(3.8)

Now, for the large-scale dynamical system $\mathcal{G}$, define the input $u(t) \triangleq S(t)$ and the output $y(t) \triangleq d(E(t))$. Hence, it follows from (3.8) that the large-scale dynamical system $\mathcal{G}$ is *lossless* [100] with respect to the *energy supply rate* $r(u, y) \triangleq \mathbf{e}^T u - \mathbf{e}^T y$ and with the *energy storage function* $U(E) \triangleq \mathbf{e}^T E$, $E \in \overline{\mathbb{R}}_+^q$. The following lemma is required for our next result.

Lemma 3.1 *Consider the large-scale dynamical system* $\mathcal{G}$ *with power balance equation (3.5). Then for every equilibrium state* $E_e \in \overline{\mathbb{R}}_+^q$ *and every* $\varepsilon > 0$ *and* $T > 0$, *there exist* $S_e \in \mathbb{R}^q$, $\alpha > 0$, *and* $\hat{T} \in [0, T]$ *such that for every* $\hat{E} \in \overline{\mathbb{R}}_+^q$ *with* $\|\hat{E} - E_e\| \leq \alpha T$, *there exists* $S : [0, \hat{T}] \to \mathbb{R}^q$ *such that* $\|S(t) - S_e\| \leq \varepsilon$, $t \in [0, \hat{T}]$, *and* $E(t) = E_e + \frac{(\hat{E} - E_e)}{\hat{T}} t$, $t \in [0, \hat{T}]$.

Proof. Note that with $S_e \triangleq d(E_e) - w(E_e)$, the state $E_e \in \overline{\mathbb{R}}_+^q$ is an equilibrium state of (3.5). Let $\theta > 0$ and $T > 0$, and define

$$M(\theta, T) \triangleq \sup_{E \in \overline{\mathcal{B}}_1(0), \, t \in [0, T]} \|w(E_e + \theta t E) - d(E_e + \theta t E) + S_e\|. \quad (3.9)$$

Note that for every $T > 0$, $\lim_{\theta \to 0^+} M(\theta, T) = 0$, and for every $\theta > 0$, $\lim_{T \to 0^+} M(\theta, T) = 0$. Next, let $\varepsilon > 0$ and $T > 0$ be given, and let $\alpha > 0$ be such that $M(\alpha, T) + \alpha \leq \varepsilon$. (The existence of such an α is guaranteed since $M(\alpha, T) \to 0$ as $\alpha \to 0^+$.) Now, let $\hat{E} \in \overline{\mathbb{R}}_+^q$ be such that $\|\hat{E} - E_e\| \le \alpha T$. With $\hat{T} \triangleq \frac{\|\hat{E} - E_e\|}{\alpha} \le T$ and

$$S(t) = -w(E(t)) + d(E(t)) + \alpha \frac{(\hat{E} - E_e)}{\|\hat{E} - E_e\|}, \quad t \in [0, \hat{T}], \quad (3.10)$$

it follows that

$$E(t) = E_e + \frac{(\hat{E} - E_e)}{\|\hat{E} - E_e\|} \alpha t, \quad t \in [0, \hat{T}], \quad\quad\quad (3.11)$$

is a solution to (3.5). The result is now immediate by noting that $E(\hat{T}) = \hat{E}$ and

$$\|S(t) - S_e\| \le \left\| w\left(E_e + \frac{(\hat{E} - E_e)}{\|\hat{E} - E_e\|} \alpha t\right) - d\left(E_e + \frac{(\hat{E} - E_e)}{\|\hat{E} - E_e\|} \alpha t\right) + S_e \right\| + \alpha$$
$$\le M(\alpha, T) + \alpha$$
$$\le \varepsilon, \quad t \in [0, \hat{T}]. \quad\quad\quad (3.12)$$

$\square$

It follows from Lemma 3.1 that the large-scale dynamical system $\mathcal{G}$ with the power balance equation (3.5) is *reachable* from and *controllable* to the origin in $\overline{\mathbb{R}}_+^q$. Recall that the large-scale dynamical system $\mathcal{G}$ with the power balance equation (3.5) is reachable from the origin in $\overline{\mathbb{R}}_+^q$ if, for all $E_0 = E(t_0) \in \overline{\mathbb{R}}_+^q$, there exists a finite time $t_i \le t_0$ and a square integrable input $S(\cdot)$ defined on $[t_i, t_0]$ such that the state $E(t)$, $t \ge t_i$, can be driven from $E(t_i) = 0$ to $E(t_0) = E_0$. Alternatively, $\mathcal{G}$ is controllable to the origin in $\overline{\mathbb{R}}_+^q$ if, for all $E_0 = E(t_0) \in \overline{\mathbb{R}}_+^q$, there exists a finite time $t_f \ge t_0$ and a square integrable input $S(\cdot)$ defined on $[t_0, t_f]$ such that the state $E(t)$, $t \ge t_0$, can be driven from $E(t_0) = E_0$ to $E(t_f) = 0$. We let $\mathcal{U}_r$ denote the set of all bounded continuous power inputs (heat fluxes) to the large-scale dynamical system $\mathcal{G}$ such that for any $T \ge -t_0$ the system energy state can be driven from $E(-T) = 0$ to $E(t_0) = E_0 \in \overline{\mathbb{R}}_+^q$ by $S(\cdot) \in \mathcal{U}_r$, and we let $\mathcal{U}_c$ denote the set of all bounded continuous power inputs (heat fluxes) to the large-scale dynamical system $\mathcal{G}$ such that for any $T \ge t_0$ the system energy state can be driven from $E(t_0) = E_0 \in \overline{\mathbb{R}}_+^q$ to $E(T) = 0$ by $S(\cdot) \in \mathcal{U}_c$. Furthermore, let $\mathcal{U}$ be an input space that is a subset of bounded continuous $\mathbb{R}^q$-valued functions on $\mathbb{R}$. The spaces $\mathcal{U}_r$, $\mathcal{U}_c$, and $\mathcal{U}$ are assumed to be closed under the shift operator, that is, if $S(\cdot) \in \mathcal{U}$ (respectively, $\mathcal{U}_c$ or $\mathcal{U}_r$), then the function S_T defined by $S_T(t) \triangleq S(t + T)$ is contained in $\mathcal{U}$ (respectively, $\mathcal{U}_c$ or $\mathcal{U}_r$) for all $T \ge 0$.

The next result establishes the uniqueness of the internal energy function $U(E)$, $E \in \overline{\mathbb{R}}_+^q$, for our large-scale dynamical system $\mathcal{G}$. For this result define the *available energy* of the large-scale dynamical system $\mathcal{G}$ by

$$U_{\mathrm{a}}(E_0) \triangleq - \inf_{u(\cdot) \in \mathcal{U}, T \geq t_0} \int_{t_0}^{T} [\mathbf{e}^{\mathrm{T}} u(t) - \mathbf{e}^{\mathrm{T}} y(t)] \mathrm{d}t, \quad E_0 \in \overline{\mathbb{R}}_+^q, \quad (3.13)$$

and the *required energy supply* of the large-scale dynamical system $\mathcal{G}$ by

$$U_{\mathrm{r}}(E_0) \triangleq \inf_{u(\cdot) \in \mathcal{U}_{\mathrm{r}}, T \geq -t_0} \int_{-T}^{t_0} [\mathbf{e}^{\mathrm{T}} u(t) - \mathbf{e}^{\mathrm{T}} y(t)] \mathrm{d}t, \quad E_0 \in \overline{\mathbb{R}}_+^q. \quad (3.14)$$

Note that the available energy $U_{\mathrm{a}}(E)$ is the maximum amount of stored energy (net heat) that can be extracted from the large-scale dynamical system $\mathcal{G}$ at any time T, and the required energy supply $U_{\mathrm{r}}(E)$ is the minimum amount of energy (net heat) that can be delivered to the large-scale dynamical system $\mathcal{G}$ to transfer it from a state of minimum potential $E(-T) = 0$ to a given state $E(t_0) = E_0$.

Theorem 3.1 *Consider the large-scale dynamical system $\mathcal{G}$ with power balance equation given by (3.5). Then $\mathcal{G}$ is lossless with respect to the energy supply rate $r(u, y) = \mathbf{e}^{\mathrm{T}} u - \mathbf{e}^{\mathrm{T}} y$, where $u(t) \equiv S(t)$ and $y(t) \equiv d(E(t))$, and with the unique energy storage function corresponding to the total energy of the system $\mathcal{G}$ given by*

$$\begin{aligned} U(E_0) &= \mathbf{e}^{\mathrm{T}} E_0 \\ &= -\int_{t_0}^{T_+} [\mathbf{e}^{\mathrm{T}} u(t) - \mathbf{e}^{\mathrm{T}} y(t)] \mathrm{d}t \\ &= \int_{-T_-}^{t_0} [\mathbf{e}^{\mathrm{T}} u(t) - \mathbf{e}^{\mathrm{T}} y(t)] \mathrm{d}t, \quad E_0 \in \overline{\mathbb{R}}_+^q, \quad (3.15) \end{aligned}$$

where $E(t)$, $t \geq t_0$, is the solution to (3.5) with admissible input $u(\cdot) \in \mathcal{U}$, $E(-T_-) = 0$, $E(T_+) = 0$, and $E(t_0) = E_0 \in \overline{\mathbb{R}}_+^q$. Furthermore,

$$0 \leq U_{\mathrm{a}}(E_0) = U(E_0) = U_{\mathrm{r}}(E_0) < \infty, \quad E_0 \in \overline{\mathbb{R}}_+^q. \quad (3.16)$$

Proof. Note that it follows from (3.8) that $\mathcal{G}$ is lossless with respect to the energy supply rate $r(u, y) = \mathbf{e}^{\mathrm{T}} u - \mathbf{e}^{\mathrm{T}} y$ and with the energy storage function $U(E) = \mathbf{e}^{\mathrm{T}} E$, $E \in \overline{\mathbb{R}}_+^q$. Since, by Lemma 3.1, $\mathcal{G}$ is reachable from and controllable to the origin in $\overline{\mathbb{R}}_+^q$, it follows from

(3.8), with $E(t_0) = E_0 \in \overline{\mathbb{R}}_+^q$ and $E(T_+) = 0$ for some $T_+ \geq t_0$ and $u(\cdot) \in \mathcal{U}$, that

$$
\begin{aligned}
\mathbf{e}^{\mathrm{T}} E_0 &= -\int_{t_0}^{T_+} [\mathbf{e}^{\mathrm{T}} u(t) - \mathbf{e}^{\mathrm{T}} y(t)] \mathrm{d}t \\
&\leq \sup_{u(\cdot) \in \mathcal{U}, T \geq t_0} \left[-\int_{t_0}^{T} [\mathbf{e}^{\mathrm{T}} u(t) - \mathbf{e}^{\mathrm{T}} y(t)] \mathrm{d}t \right] \\
&= -\inf_{u(\cdot) \in \mathcal{U}, T \geq t_0} \int_{t_0}^{T} [\mathbf{e}^{\mathrm{T}} u(t) - \mathbf{e}^{\mathrm{T}} y(t)] \mathrm{d}t \\
&= U_{\mathrm{a}}(E_0), \quad E_0 \in \overline{\mathbb{R}}_+^q.
\end{aligned}
\tag{3.17}
$$

Alternatively, it follows from (3.8), with $E(-T_-) = 0$ for some $-T_- \leq t_0$ and $u(\cdot) \in \mathcal{U}_{\mathrm{r}}$, that

$$
\begin{aligned}
\mathbf{e}^{\mathrm{T}} E_0 &= \int_{-T_-}^{t_0} [\mathbf{e}^{\mathrm{T}} u(t) - \mathbf{e}^{\mathrm{T}} y(t)] \mathrm{d}t \\
&\geq \inf_{u(\cdot) \in \mathcal{U}_{\mathrm{r}}, T \geq -t_0} \int_{-T}^{t_0} [\mathbf{e}^{\mathrm{T}} u(t) - \mathbf{e}^{\mathrm{T}} y(t)] \mathrm{d}t \\
&= U_{\mathrm{r}}(E_0), \quad E_0 \in \overline{\mathbb{R}}_+^q.
\end{aligned}
\tag{3.18}
$$

Thus, (3.17) and (3.18) imply that (3.15) is satisfied and

$$
U_{\mathrm{r}}(E_0) \leq \mathbf{e}^{\mathrm{T}} E_0 \leq U_{\mathrm{a}}(E_0), \quad E_0 \in \overline{\mathbb{R}}_+^q.
\tag{3.19}
$$

Conversely, it follows from (3.8) and the fact that $U(E) = \mathbf{e}^{\mathrm{T}} E \geq 0$, $E \in \overline{\mathbb{R}}_+^q$, that, for all $T \geq t_0$ and $u(\cdot) \in \mathcal{U}$,

$$
\mathbf{e}^{\mathrm{T}} E(t_0) \geq -\int_{t_0}^{T} [\mathbf{e}^{\mathrm{T}} u(t) - \mathbf{e}^{\mathrm{T}} y(t)] \mathrm{d}t, \quad E(t_0) \in \overline{\mathbb{R}}_+^q,
\tag{3.20}
$$

which implies that

$$
\begin{aligned}
\mathbf{e}^{\mathrm{T}} E(t_0) &\geq \sup_{u(\cdot) \in \mathcal{U}, T \geq t_0} \left[-\int_{t_0}^{T} [\mathbf{e}^{\mathrm{T}} u(t) - \mathbf{e}^{\mathrm{T}} y(t)] \mathrm{d}t \right] \\
&= -\inf_{u(\cdot) \in \mathcal{U}, T \geq t_0} \int_{t_0}^{T} [\mathbf{e}^{\mathrm{T}} u(t) - \mathbf{e}^{\mathrm{T}} y(t)] \mathrm{d}t \\
&= U_{\mathrm{a}}(E(t_0)), \quad E(t_0) \in \overline{\mathbb{R}}_+^q.
\end{aligned}
\tag{3.21}
$$

Furthermore, it follows from the definition of $U_{\mathrm{a}}(\cdot)$ that $U_{\mathrm{a}}(E) \geq 0$, $E \in \overline{\mathbb{R}}_+^q$, since the infimum in (3.13) is taken over the set of values containing the zero value ($T = t_0$). Next, note that it follows from

(3.8), with $E(t_0) \in \overline{\mathbb{R}}_+^q$ and $E(-T) = 0$ for all $T \geq -t_0$ and $u(\cdot) \in \mathcal{U}_r$, that

$$\mathbf{e}^T E(t_0) = \int_{-T}^{t_0} [\mathbf{e}^T u(t) - \mathbf{e}^T y(t)] dt$$

$$= \inf_{u(\cdot) \in \mathcal{U}_r, T \geq -t_0} \int_{-T}^{t_0} [\mathbf{e}^T u(t) - \mathbf{e}^T y(t)] dt$$

$$= U_r(E(t_0)), \quad E(t_0) \in \overline{\mathbb{R}}_+^q. \tag{3.22}$$

Moreover, since the system $\mathcal{G}$ is reachable from the origin, it follows that for every $E(t_0) \in \overline{\mathbb{R}}_+^q$, there exists $T \geq -t_0$ and $u(\cdot) \in \mathcal{U}_r$ such that

$$\int_{-T}^{t_0} (\mathbf{e}^T u(t) - \mathbf{e}^T y(t)) dt \tag{3.23}$$

is finite, and hence, $U_r(E(t_0)) < \infty$, $E(t_0) \in \overline{\mathbb{R}}_+^q$. Finally, combining (3.19), (3.21), and (3.22), it follows that (3.16) holds. □

It follows from (3.16) and the definitions of available energy $U_a(E_0)$ and the required energy supply $U_r(E_0)$, $E_0 \in \overline{\mathbb{R}}_+^q$, that the large-scale dynamical system $\mathcal{G}$ can deliver to its surroundings all of its stored subsystem energies and can store all of the work done to all of its subsystems. This is in essence a statement of the first law of thermodynamics and places no limitation on the possibility of transforming heat into work or work into heat. In the case where $S(t) \equiv 0$, it follows from (3.8) and the fact that $\sigma_{ii}(E) \geq 0$, $E \in \overline{\mathbb{R}}_+^q$, $i = 1, ..., q$, that the zero solution $E(t) \equiv 0$ of the large-scale dynamical system $\mathcal{G}$ with the power balance equation (3.5) is Lyapunov stable with Lyapunov function $U(E)$ corresponding to the total energy in the system.

3.3 Entropy and the Second Law of Thermodynamics

The nonlinear power balance equation (3.5) can exhibit a full range of nonlinear behavior, including bifurcations, limit cycles, and even chaos. However, a thermodynamically consistent energy flow model should ensure that the evolution of the system energy is diffusive (parabolic) in character with convergent subsystem energies. As established in Section 2.4, such a system model would guarantee the absence of the *Poincaré recurrence* phenomenon [4], which states that every finite-dimensional, *isolated* (i.e., $S(t) \equiv 0$ and $d(E) \equiv 0$) dynamical system with *volume-preserving*[1] trajectories (subsystem energies) will return arbitrarily close to its initial system state (energy) infinitely many times. This of course would violate the second law of thermodynamics, since subsystem energies (temperatures) would be allowed to return to their starting state and thereby subverting the diffusive character of the dynamical system. Hence, to ensure a thermodynamically consistent energy flow model, we require the following axioms.[2] For the statement of these axioms, we first recall the following graph-theoretic notions.

Definition 3.1 ([6]) *A directed graph $G(\mathcal{C})$ associated with the connectivity matrix $\mathcal{C} \in \mathbb{R}^{q \times q}$ has vertices $\{1, 2, ..., q\}$ and an arc from vertex i to vertex j, $i \neq j$, if and only if $\mathcal{C}_{(j,i)} \neq 0$. A graph $G(\mathcal{C})$ associated with the connectivity matrix $\mathcal{C} \in \mathbb{R}^{q \times q}$ is a directed graph for which the arc set is symmetric, that is, $\mathcal{C} = \mathcal{C}^{\mathrm{T}}$. We say that $G(\mathcal{C})$ is* strongly connected *if for any ordered pair of vertices (i, j), $i \neq j$, there exists a path (i.e., a sequence of arcs) leading from i to j.*

Recall that the connectivity matrix $\mathcal{C} \in \mathbb{R}^{q \times q}$ is *irreducible*, that is, there does not exist a permutation matrix such that $\mathcal{C}$ is cogredient to a lower-block triangular matrix, if and only if $G(\mathcal{C})$ is strongly connected (see Theorem 2.7 of [6]). Let $\phi_{ij}(E) \triangleq \sigma_{ij}(E) - \sigma_{ji}(E)$, $E \in \overline{\mathbb{R}}_+^q$, denote the net energy flow from the jth subsystem $\mathcal{G}_j$ to the ith subsystem $\mathcal{G}_i$ of the large-scale dynamical system $\mathcal{G}$.

Axiom *i*) *For the connectivity matrix $\mathcal{C} \in \mathbb{R}^{q \times q}$ associated with the large-scale dynamical system $\mathcal{G}$ defined by*

$$\mathcal{C}_{(i,j)} \triangleq \begin{cases} 0, & \text{if } \phi_{ij}(E) \equiv 0, \\ 1, & \text{otherwise,} \end{cases} \quad i \neq j, \quad i, j = 1, ..., q, \quad (3.24)$$

and

$$\mathcal{C}_{(i,i)} \triangleq - \sum_{k=1, \, k \neq i}^{q} \mathcal{C}_{(k,i)}, \quad i = j, \quad i = 1, ..., q, \quad (3.25)$$

---

[1]A dynamical system is *volume-preserving* if the volume of an arbitrary region of the state space is conserved by the time evolution of the system, even though the shape of the region may change dramatically.

[2]It can be argued here that a more appropriate terminology is *assumptions* rather than *axioms* since, as will be seen, these are statements taken to be true and used as premises in order to infer certain results, but may not otherwise be accepted. However, as we will see, these statements are equivalent (within our formulation) to the stipulated postulates of the zeroth and second laws of thermodynamics involving transitivity of a thermal equilibrium and heat flowing from *hotter* to *colder* bodies, and as such we refer to them as *axioms*.

rank $\mathcal{C} = q - 1$, *and for* $\mathcal{C}_{(i,j)} = 1$, $i \neq j$, $\phi_{ij}(E) = 0$ *if and only if* $E_i = E_j$.

Axiom ii) *For* $i, j = 1, ..., q$, $(E_i - E_j)\phi_{ij}(E) \leq 0$, $E \in \overline{\mathbb{R}}_+^q$.

The fact that $\phi_{ij}(E) = 0$ if and only if $E_i = E_j$, $i \neq j$, implies that subsystems $\mathcal{G}_i$ and $\mathcal{G}_j$ of $\mathcal{G}$ are *connected*; alternatively, $\phi_{ij}(E) \equiv 0$ implies that $\mathcal{G}_i$ and $\mathcal{G}_j$ are *disconnected*. Axiom i) implies that if the energies in the connected subsystems $\mathcal{G}_i$ and $\mathcal{G}_j$ are equal, then energy exchange between these subsystems is not possible. This statement is consistent with the *zeroth law of thermodynamics*, which postulates that temperature equality is a necessary and sufficient condition for thermal equilibrium. Furthermore, it follows from the fact that $\mathcal{C} = \mathcal{C}^{\mathrm{T}}$ and rank $\mathcal{C} = q - 1$ that the connectivity matrix $\mathcal{C}$ is irreducible, which implies that for any pair of subsystems $\mathcal{G}_i$ and $\mathcal{G}_j$, $i \neq j$, of $\mathcal{G}$ there exists a sequence of connectors (arcs) of $\mathcal{G}$ that connect $\mathcal{G}_i$ and $\mathcal{G}_j$. Axiom ii) implies that energy flows from more energetic subsystems to less energetic subsystems and is consistent with the *second law of thermodynamics*, which states that heat (energy) must flow in the direction of lower temperatures.[3] Furthermore, note that $\phi_{ij}(E) = -\phi_{ji}(E)$, $E \in \overline{\mathbb{R}}_+^q$, $i \neq j$, $i, j = 1, ..., q$, which implies conservation of energy between lossless subsystems. With $S(t) \equiv 0$, Axioms i) and ii) along with the fact that $\phi_{ij}(E) = -\phi_{ji}(E)$, $E \in \overline{\mathbb{R}}_+^q$, $i \neq j$, $i, j = 1, ..., q$, imply that at a given instant of time, energy can only be transported, stored, or dissipated but not created, and the maximum amount of energy that can be transported and/or dissipated from a subsystem cannot exceed the energy in the subsystem.

Next, we show that the classical Clausius equality and inequality for reversible and irreversible thermodynamics over cyclic motions are satisfied for our thermodynamically consistent energy flow model. For this result $\oint$ denotes a cyclic integral evaluated along an arbitrary closed path of (3.5) in $\overline{\mathbb{R}}_+^q$; that is, $\oint \triangleq \int_{t_0}^{t_f}$ with $t_f \geq t_0$ and $S(\cdot) \in \mathcal{U}$ such that $E(t_f) = E(t_0) = E_0 \in \overline{\mathbb{R}}_+^q$.

Proposition 3.2 *Consider the large-scale dynamical system $\mathcal{G}$ with power balance equation (3.5), and assume that Axioms i) and ii) hold. Then for all $E_0 \in \overline{\mathbb{R}}_+^q$, $t_f \geq t_0$, and $S(\cdot) \in \mathcal{U}$ such that $E(t_f) = E(t_0) =*

---

[3]It is important to note that our formulation of the second law of thermodynamics as given by Axiom ii) does not require the mentioning of temperature nor the more primitive subjective notions of hotness or coldness. As we will see later, temperature is defined in terms of the system entropy after we establish the existence of a unique, continuously differentiable entropy function for $\mathcal{G}$.

E_0,

$$\int_{t_0}^{t_f} \sum_{i=1}^{q} \frac{S_i(t) - \sigma_{ii}(E(t))}{c + E_i(t)} dt = \oint \sum_{i=1}^{q} \frac{dQ_i(t)}{c + E_i(t)} \leq 0, \quad (3.26)$$

where $c > 0$, $dQ_i(t) \triangleq [S_i(t) - \sigma_{ii}(E(t))]dt$, $i = 1, ..., q$, is the amount of net energy (heat) received by the ith subsystem over the infinitesimal time interval dt, and $E(t)$, $t \geq t_0$, is the solution to (3.5) with initial condition $E(t_0) = E_0$. Furthermore,

$$\oint \sum_{i=1}^{q} \frac{dQ_i(t)}{c + E_i(t)} = 0 \qquad (3.27)$$

if and only if there exists a continuous function $\alpha : [t_0, t_f] \to \overline{\mathbb{R}}_+$ such that $E(t) = \alpha(t)\mathbf{e}$, $t \in [t_0, t_f]$.

Proof. Since, by Proposition 3.1, $E(t) \geq\geq 0$, $t \geq t_0$, and $\phi_{ij}(E) = -\phi_{ji}(E)$, $E \in \overline{\mathbb{R}}_+^q$, $i \neq j$, $i, j = 1, ..., q$, it follows from (3.5) and Axiom $ii)$ that

$$\oint \sum_{i=1}^{q} \frac{dQ_i(t)}{c + E_i(t)} = \int_{t_0}^{t_f} \sum_{i=1}^{q} \frac{\dot{E}_i(t) - \sum_{j=1, j\neq i}^{q} \phi_{ij}(E(t))}{c + E_i(t)} dt$$

$$= \sum_{i=1}^{q} \log_e \left(\frac{c + E_i(t_f)}{c + E_i(t_0)} \right) - \int_{t_0}^{t_f} \sum_{i=1}^{q} \sum_{j=1, j\neq i}^{q} \frac{\phi_{ij}(E(t))}{c + E_i(t)} dt$$

$$= -\int_{t_0}^{t_f} \sum_{i=1}^{q} \sum_{j=i+1}^{q} \left(\frac{\phi_{ij}(E(t))}{c + E_i(t)} - \frac{\phi_{ij}(E(t))}{c + E_j(t)} \right) dt$$

$$= -\int_{t_0}^{t_f} \sum_{i=1}^{q} \sum_{j=i+1}^{q} \frac{\phi_{ij}(E(t))(E_j(t) - E_i(t))}{(c + E_i(t))(c + E_j(t))} dt$$

$$\leq 0, \qquad (3.28)$$

which proves (3.26).

To show (3.27), note that it follows from (3.28), Axiom $i)$, and Axiom $ii)$ that (3.27) holds if and only if $E_i(t) = E_j(t)$, $t \in [t_0, t_f]$, $i \neq j$, $i, j = 1, ..., q$, or, equivalently, there exists a continuous function $\alpha : [t_0, t_f] \to \overline{\mathbb{R}}_+$ such that $E(t) = \alpha(t)\mathbf{e}$, $t \in [t_0, t_f]$. $\square$

Inequality (3.26) is a generalization of Clausius' inequality for reversible and irreversible thermodynamics as applied to large-scale dynamical systems and restricts the manner in which the system dissipates (scaled) heat over cyclic motions. It follows from Axiom

i) and (3.5) that for the *adiabatically isolated* large-scale dynamical system $\mathcal{G}$ (that is, $S(t) \equiv 0$ and $d(E(t)) \equiv 0$), the energy states given by $E_{\mathrm{e}} = \alpha \mathbf{e}$, $\alpha \geq 0$, correspond to the equilibrium energy states of $\mathcal{G}$. Thus, as in classical thermodynamics, we can define an *equilibrium process* as a process in which the trajectory of the large-scale dynamical system $\mathcal{G}$ moves along the equilibrium manifold $\mathcal{M}_{\mathrm{e}} \triangleq \{E \in \overline{\mathbb{R}}_+^q : E = \alpha \mathbf{e}, \alpha \geq 0\}$ corresponding to the set of equilibria of the isolated[4] system $\mathcal{G}$. The power input that can generate such a trajectory can be given by $S(t) = d(E(t)) + u(t)$, $t \geq t_0$, where $u(\cdot) \in \mathcal{U}$ is such that $u_i(t) \equiv u_j(t)$, $i \neq j$, $i, j = 1, ..., q$. Our definition of an equilibrium transformation involves a continuous succession of intermediate states that differ by infinitesimals from equilibrium system states and thus can only connect initial and final states, which are states of equilibrium. This process need not be slowly varying, and hence, equilibrium and quasistatic processes are not synonymous in this monograph. Alternatively, a *nonequilibrium process* is a process that does not lie on the equilibrium manifold $\mathcal{M}_{\mathrm{e}}$. Hence, it follows from Axiom i) that for an equilibrium process $\phi_{ij}(E(t)) = 0$, $t \geq t_0$, $i \neq j$, $i, j = 1, ..., q$, and thus, by Proposition 3.2, inequality (3.26) is satisfied as an equality. Alternatively, for a nonequilibrium process it follows from Axioms i) and ii) that (3.26) is satisfied as a strict inequality.

Next, we give a deterministic definition of entropy for the large-scale dynamical system $\mathcal{G}$ that is consistent with the classical thermodynamic definition of entropy.

Definition 3.2 *For the large-scale dynamical system $\mathcal{G}$ with power balance equation (3.5), a function $\mathcal{S} : \overline{\mathbb{R}}_+^q \to \mathbb{R}$ satisfying*

$$\mathcal{S}(E(t_2)) \geq \mathcal{S}(E(t_1)) + \int_{t_1}^{t_2} \sum_{i=1}^{q} \frac{S_i(t) - \sigma_{ii}(E(t))}{c + E_i(t)} dt \qquad (3.29)$$

for any $t_2 \geq t_1 \geq t_0$ and $S(\cdot) \in \mathcal{U}$ is called the entropy *function of $\mathcal{G}$.*

Next, we show that (3.26) guarantees the existence of an entropy function for $\mathcal{G}$. For this result define the *available entropy* of the large-scale dynamical system $\mathcal{G}$ by

$$\mathcal{S}_{\mathrm{a}}(E_0) \triangleq - \sup_{S(\cdot) \in \mathcal{U}_{\mathrm{c}}, \, T \geq t_0} \int_{t_0}^{T} \sum_{i=1}^{q} \frac{S_i(t) - \sigma_{ii}(E(t))}{c + E_i(t)} dt, \qquad (3.30)$$

---

[4]Since in this section we are not considering work performed by and on the system, the notions of an *isolated* system and an *adiabatically isolated* system are equivalent.

where $E(t_0) = E_0 \in \overline{\mathbb{R}}_+^q$ and $E(T) = 0$, and define the *required entropy supply* of the large-scale dynamical system $\mathcal{G}$ by

$$\mathcal{S}_{\mathrm{r}}(E_0) \triangleq \sup_{S(\cdot) \in \mathcal{U}_{\mathrm{r}}, T \geq -t_0} \int_{-T}^{t_0} \sum_{i=1}^{q} \frac{S_i(t) - \sigma_{ii}(E(t))}{c + E_i(t)} \mathrm{d}t, \quad (3.31)$$

where $E(-T) = 0$ and $E(t_0) = E_0 \in \overline{\mathbb{R}}_+^q$. Note that the available entropy $\mathcal{S}_{\mathrm{a}}(E_0)$ is the minimum amount of scaled heat (entropy) that can be extracted from the large-scale dynamical system $\mathcal{G}$ in order to transfer it from an initial state $E(t_0) = E_0$ to $E(T) = 0$. Alternatively, the required entropy supply $\mathcal{S}_{\mathrm{r}}(E_0)$ is the maximum amount of scaled heat (entropy) that can be delivered to $\mathcal{G}$ to transfer it from the origin to a given initial state $E(t_0) = E_0$.

Theorem 3.2 *Consider the large-scale dynamical system $\mathcal{G}$ with power balance equation (3.5), and assume that Axiom ii) holds. Then there exists an entropy function for $\mathcal{G}$. Moreover, $\mathcal{S}_{\mathrm{a}}(E)$, $E \in \overline{\mathbb{R}}_+^q$, and $\mathcal{S}_{\mathrm{r}}(E)$, $E \in \overline{\mathbb{R}}_+^q$, are possible entropy functions for $\mathcal{G}$ with $\mathcal{S}_{\mathrm{a}}(0) = \mathcal{S}_{\mathrm{r}}(0) = 0$. Finally, all entropy functions $\mathcal{S}(E)$, $E \in \overline{\mathbb{R}}_+^q$, for $\mathcal{G}$ satisfy*

$$\mathcal{S}_{\mathrm{r}}(E) \leq \mathcal{S}(E) - \mathcal{S}(0) \leq \mathcal{S}_{\mathrm{a}}(E), \quad E \in \overline{\mathbb{R}}_+^q. \quad (3.32)$$

Proof. Since, by Lemma 3.1, $\mathcal{G}$ is controllable to and reachable from the origin in $\overline{\mathbb{R}}_+^q$, it follows from (3.30) and (3.31) that $\mathcal{S}_{\mathrm{a}}(E_0) < \infty$, $E_0 \in \overline{\mathbb{R}}_+^q$, and $\mathcal{S}_{\mathrm{r}}(E_0) > -\infty$, $E_0 \in \overline{\mathbb{R}}_+^q$, respectively. Next, let $E_0 \in \overline{\mathbb{R}}_+^q$, and let $S(\cdot) \in \mathcal{U}$ be such that $E(t_{\mathrm{i}}) = E(t_{\mathrm{f}}) = 0$ and $E(t_0) = E_0$, where $t_{\mathrm{i}} < t_0 < t_{\mathrm{f}}$. In this case, it follows from (3.26) that

$$\int_{t_{\mathrm{i}}}^{t_{\mathrm{f}}} \sum_{i=1}^{q} \frac{S_i(t) - \sigma_{ii}(E(t))}{c + E_i(t)} \mathrm{d}t \leq 0 \quad (3.33)$$

or, equivalently,

$$\int_{t_{\mathrm{i}}}^{t_0} \sum_{i=1}^{q} \frac{S_i(t) - \sigma_{ii}(E(t))}{c + E_i(t)} \mathrm{d}t \leq -\int_{t_0}^{t_{\mathrm{f}}} \sum_{i=1}^{q} \frac{S_i(t) - \sigma_{ii}(E(t))}{c + E_i(t)} \mathrm{d}t. \quad (3.34)$$

Now, taking the supremum on both sides of (3.34) over all $S(\cdot) \in \mathcal{U}_{\mathrm{r}}$ and $t_{\mathrm{i}} \leq t_0$ yields

$$\mathcal{S}_{\mathrm{r}}(E_0) = \sup_{S(\cdot) \in \mathcal{U}_{\mathrm{r}}, t_{\mathrm{i}} \leq t_0} \int_{t_{\mathrm{i}}}^{t_0} \sum_{i=1}^{q} \frac{S_i(t) - \sigma_{ii}(E(t))}{c + E_i(t)} \mathrm{d}t$$

$$\leq -\int_{t_0}^{t_{\mathrm{f}}} \sum_{i=1}^{q} \frac{S_i(t) - \sigma_{ii}(E(t))}{c + E_i(t)} \mathrm{d}t. \quad (3.35)$$

Next, taking the infimum on both sides of (3.35) over all $S(\cdot) \in \mathcal{U}_c$ and $t_f \geq t_0$, we obtain $S_r(E_0) \leq S_a(E_0)$, $E_0 \in \overline{\mathbb{R}}_+^q$, which implies that $-\infty < S_r(E_0) \leq S_a(E_0) < \infty$, $E_0 \in \overline{\mathbb{R}}_+^q$. Hence, the functions $S_a(\cdot)$ and $S_r(\cdot)$ are well defined.

Next, it follows from the definition of $S_a(\cdot)$ that for any $T \geq t_1$ and $S(\cdot) \in \mathcal{U}_c$ such that $E(t_1) \in \overline{\mathbb{R}}_+^q$ and $E(T) = 0$,

$$-S_a(E(t_1)) \geq \int_{t_1}^{t_2} \sum_{i=1}^{q} \frac{S_i(t) - \sigma_{ii}(E(t))}{c + E_i(t)} dt$$

$$+ \int_{t_2}^{T} \sum_{i=1}^{q} \frac{S_i(t) - \sigma_{ii}(E(t))}{c + E_i(t)} dt, \quad t_1 \leq t_2 \leq T, \quad (3.36)$$

and hence,

$$-S_a(E(t_1)) \geq \int_{t_1}^{t_2} \sum_{i=1}^{q} \frac{S_i(t) - \sigma_{ii}(E(t))}{c + E_i(t)} dt$$

$$+ \sup_{S(\cdot) \in \mathcal{U}_c, T \geq t_2} \int_{t_2}^{T} \sum_{i=1}^{q} \frac{S_i(t) - \sigma_{ii}(E(t))}{c + E_i(t)} dt$$

$$= \int_{t_1}^{t_2} \sum_{i=1}^{q} \frac{S_i(t) - \sigma_{ii}(E(t))}{c + E_i(t)} dt - S_a(E(t_2)), \quad (3.37)$$

which implies that $S_a(E)$, $E \in \overline{\mathbb{R}}_+^q$, satisfies (3.29). Thus, $S_a(E)$, $E \in \overline{\mathbb{R}}_+^q$, is a possible entropy function for $\mathcal{G}$. Note that with $E(t_0) = E(T) = 0$ it follows from (3.26) that the supremum in (3.30) is taken over the set of negative semi-definite values with one of the values being zero for $S(t) \equiv 0$. Thus, $S_a(0) = 0$.

Similarly, it follows from the definition of $S_r(\cdot)$ that for any $T \geq -t_2$ and $S(\cdot) \in \mathcal{U}_r$ such that $E(t_2) \in \overline{\mathbb{R}}_+^q$ and $E(-T) = 0$,

$$S_r(E(t_2)) \geq \int_{-T}^{t_1} \sum_{i=1}^{q} \frac{S_i(t) - \sigma_{ii}(E(t))}{c + E_i(t)} dt$$

$$+ \int_{t_1}^{t_2} \sum_{i=1}^{q} \frac{S_i(t) - \sigma_{ii}(E(t))}{c + E_i(t)} dt, \quad -T \leq t_1 \leq t_2, \quad (3.38)$$

and hence,

$$S_r(E(t_2)) \geq \int_{t_1}^{t_2} \sum_{i=1}^{q} \frac{S_i(t) - \sigma_{ii}(E(t))}{c + E_i(t)} dt$$

$$+ \sup_{S(\cdot) \in \mathcal{U}_r, T \geq -t_1} \int_{-T}^{t_1} \sum_{i=1}^{q} \frac{S_i(t) - \sigma_{ii}(E(t))}{c + E_i(t)} dt$$

$$= \int_{t_1}^{t_2} \sum_{i=1}^{q} \frac{S_i(t) - \sigma_{ii}(E(t))}{c + E_i(t)} dt + \mathcal{S}_r(E(t_1)), \quad (3.39)$$

which implies that $\mathcal{S}_r(E)$, $E \in \overline{\mathbb{R}}_+^q$, satisfies (3.29). Thus, $\mathcal{S}_r(E)$, $E \in \overline{\mathbb{R}}_+^q$, is a possible entropy function for $\mathcal{G}$. Note that with $E(t_0) = E(-T) = 0$ it follows from (3.26) that the supremum in (3.31) is taken over the set of negative semi-definite values with one of the values being zero for $S(t) \equiv 0$. Thus, $\mathcal{S}_r(0) = 0$.

Next, suppose there exists an entropy function $\mathcal{S} : \overline{\mathbb{R}}_+^q \to \mathbb{R}$ for $\mathcal{G}$, and let $E(t_2) = 0$ in (3.29). Then it follows from (3.29) that

$$\mathcal{S}(E(t_1)) - \mathcal{S}(0) \leq -\int_{t_1}^{t_2} \sum_{i=1}^{q} \frac{S_i(t) - \sigma_{ii}(E(t))}{c + E_i(t)} dt, \quad (3.40)$$

for all $t_2 \geq t_1$ and $S(\cdot) \in \mathcal{U}_c$, which implies that

$$\mathcal{S}(E(t_1)) - \mathcal{S}(0) \leq \inf_{S(\cdot) \in \mathcal{U}_c, t_2 \geq t_1} \left[-\int_{t_1}^{t_2} \sum_{i=1}^{q} \frac{S_i(t) - \sigma_{ii}(E(t))}{c + E_i(t)} dt \right]$$

$$= -\sup_{S(\cdot) \in \mathcal{U}_c, t_2 \geq t_1} \int_{t_1}^{t_2} \sum_{i=1}^{q} \frac{S_i(t) - \sigma_{ii}(E(t))}{c + E_i(t)} dt$$

$$= \mathcal{S}_a(E(t_1)). \quad (3.41)$$

Since $E(t_1)$ is arbitrary, it follows that $\mathcal{S}(E) - \mathcal{S}(0) \leq \mathcal{S}_a(E)$, $E \in \overline{\mathbb{R}}_+^q$. Alternatively, let $E(t_1) = 0$ in (3.29). Then it follows from (3.29) that

$$\mathcal{S}(E(t_2)) - \mathcal{S}(0) \geq \int_{t_1}^{t_2} \sum_{i=1}^{q} \frac{S_i(t) - \sigma_{ii}(E(t))}{c + E_i(t)} dt \quad (3.42)$$

for all $t_1 \leq t_2$ and $S(\cdot) \in \mathcal{U}_r$. Hence,

$$\mathcal{S}(E(t_2)) - \mathcal{S}(0) \geq \sup_{S(\cdot) \in \mathcal{U}_r, t_1 \leq t_2} \int_{t_1}^{t_2} \sum_{i=1}^{q} \frac{S_i(t) - \sigma_{ii}(E(t))}{c + E_i(t)} dt$$

$$= \mathcal{S}_r(E(t_2)), \quad (3.43)$$

which, since $E(t_2)$ is arbitrary, implies that $\mathcal{S}_r(E) \leq \mathcal{S}(E) - \mathcal{S}(0)$, $E \in \overline{\mathbb{R}}_+^q$. Thus, all entropy functions for $\mathcal{G}$ satisfy (3.32). $\square$

It is important to note that inequality (3.26) is equivalent to the existence of an entropy function for $\mathcal{G}$. Sufficiency is simply a statement of Theorem 3.2, while necessity follows from (3.29) with $E(t_2) = E(t_1)$. This definition of entropy leads to the second law of thermodynamics being viewed as an axiom in the context of (anti)cyclo-dissipative dynamical systems [52, 101, 102]. A similar remark holds for the definition of ectropy introduced in Section 3.4. The next result shows that all entropy functions for $\mathcal{G}$ are continuous on $\overline{\mathbb{R}}_+^q$.

Theorem 3.3 *Consider the large-scale dynamical system $\mathcal{G}$ with power balance equation (3.5), and let $S : \overline{\mathbb{R}}_+^q \to \mathbb{R}$ be an entropy function of $\mathcal{G}$. Then $S(\cdot)$ is continuous on $\overline{\mathbb{R}}_+^q$.*

Proof. Let $E_e \in \overline{\mathbb{R}}_+^q$ and $S_e \in \mathbb{R}^q$ be such that $S_e = d(E_e) - w(E_e)$. Note that with $S(t) \equiv S_e$, E_e is an equilibrium point of the power balance equation (3.5). Next, it follows from Lemma 3.1 that $\mathcal{G}$ is *locally controllable*, that is, for every $T > 0$ and $\varepsilon > 0$, the set of points that can be reached from and to E_e in time T using admissible inputs $S : [0, T] \to \mathbb{R}^q$, satisfying $\|S(t) - S_e\| < \varepsilon$, contains a neighborhood of E_e. Alternatively, this can be shown by considering the linearization of (3.5) at $E = E_e$ and $S = S_e$ given by

$$\dot{E}(t) = A(E(t) - E_e) + B(S(t) - S_e), \quad E(t_0) = E_0, \quad t \geq t_0, \quad (3.44)$$

where $A = \left. \frac{\partial w(E)}{\partial E} \right|_{E=E_e} - \left. \frac{\partial d(E)}{\partial E} \right|_{E=E_e}$ and $B = I_q$. Since $B = I_q$, it follows that

$$\text{rank} \, [B, AB, A^2 B, ..., A^{q-1} B] = q, \quad (3.45)$$

and hence, the linearized system (3.44) is controllable. Thus, it follows from Proposition 3.3 of [76] that $\mathcal{G}$ is locally controllable.

Next, let $\delta > 0$ and note that it follows from the continuity of $w(\cdot)$ and $d(\cdot)$ that there exist $T > 0$ and $\varepsilon > 0$ such that for every $S : [0, T] \to \mathbb{R}^q$ and $\|S(t) - S_e\| < \varepsilon$, $\|E(t) - E_e\| < \delta$, $t \in [0, T)$, where $S(\cdot) \in \mathcal{U}$ and $E(t)$, $t \in [0, T)$, denotes the solution to (3.5) with the initial condition E_e. Furthermore, it follows from the local controllability of $\mathcal{G}$ that for every $\hat{T} \in (0, T]$, there exists a strictly increasing, continuous function $\gamma : \mathbb{R} \to \overline{\mathbb{R}}_+^q$ such that $\gamma(0) = 0$, and for every $E_0 \in \overline{\mathbb{R}}_+^q$ such that $\|E_0 - E_e\| \leq \gamma(\hat{T})$, there exists $\hat{t} \in [0, \hat{T}]$ and an input $S : [0, \hat{T}] \to \mathbb{R}^q$ such that $\|S(t) - S_e\| < \varepsilon$, $t \in [0, \hat{t})$, and $E(\hat{t}) = E_0$. Hence, there exists $\beta > 0$ such that for every $E_0 \in \overline{\mathbb{R}}_+^q$ such that $\|E_0 - E_e\| \leq \beta$, there exists $\hat{t} \in [0, \gamma^{-1}(\|E_0 - E_e\|)]$ and an input $S : [t_0, \hat{t}] \to \mathbb{R}^q$ such that $\|S(t) - S_e\| < \varepsilon$, $t \in [0, \hat{t}]$, and $E(\hat{t}) = E_0$.

In addition, it follows from Lemma 3.1 that $S : [0, \hat{t}] \to \mathbb{R}^q$ is such that $E(t) \geq\geq 0$, $t \in [0, \hat{t}]$.

Next, since $\sigma_{ii}(\cdot)$, $i = 1, \ldots, q$, is continuous, it follows that there exists $M \in (0, \infty)$ such that

$$\sup_{\|E - E_e\| < \delta, \ \|S - S_e\| < \varepsilon} \left| \sum_{i=1}^{q} \frac{S_i - \sigma_{ii}(E)}{c + E_i} \right| = M. \qquad (3.46)$$

Hence, it follows that

$$\left| \int_0^{\hat{t}} \sum_{i=1}^{q} \frac{S_i(\sigma) - \sigma_{ii}(E(\sigma))}{c + E_i(\sigma)} \mathrm{d}\sigma \right| \leq \int_0^{\hat{t}} \left| \sum_{i=1}^{q} \frac{S_i(\sigma) - \sigma_{ii}(E(\sigma))}{c + E_i(\sigma)} \right| \mathrm{d}\sigma$$

$$\leq M\hat{t}$$

$$\leq M\gamma^{-1}(\|E_0 - E_e\|). \qquad (3.47)$$

Now, if $\mathcal{S}(\cdot)$ is an entropy function of $\mathcal{G}$, then

$$\mathcal{S}(E(\hat{t})) \geq \mathcal{S}(E_e) + \int_0^{\hat{t}} \sum_{i=1}^{q} \frac{S_i(\sigma) - \sigma_{ii}(E(\sigma))}{c + E_i(\sigma)} \mathrm{d}\sigma \qquad (3.48)$$

or, equivalently,

$$-\int_0^{\hat{t}} \sum_{i=1}^{q} \frac{S_i(\sigma) - \sigma_{ii}(E(\sigma))}{c + E_i(\sigma)} \mathrm{d}\sigma \geq \mathcal{S}(E_e) - \mathcal{S}(E(\hat{t})). \qquad (3.49)$$

If $\mathcal{S}(E_e) \geq \mathcal{S}(E(\hat{t}))$, then combining (3.47) and (3.49) yields

$$|\mathcal{S}(E_e) - \mathcal{S}(E(\hat{t}))| \leq M\gamma^{-1}(\|E_0 - E_e\|). \qquad (3.50)$$

Alternatively, if $\mathcal{S}(E(\hat{t})) \geq \mathcal{S}(E_e)$, then (3.50) can be derived by reversing the roles of E_e and $E(\hat{t})$. In particular, using the fact that $\mathcal{G}$ is locally controllable from and to E_e or, alternatively, that controllability of (3.44) is equivalent to controllability of the linearization of the time-reversed system

$$\dot{E}(t) = -A(E(t) - E_e) - B(S(t) - S_e), \quad E(t_0) = E_0, \quad t \geq t_0, (3.51)$$

similar arguments can be used to show that the set of points that can be steered in small time to E_e contains a neighborhood of $E(\hat{t})$. Hence, since $\gamma(\cdot)$ is continuous and $E(\hat{t})$ is arbitrary, it follows that $\mathcal{S}(\cdot)$ is continuous on $\overline{\mathbb{R}}_+^q$. $\qquad \square$

Next, as a direct consequence of Theorem 3.2, we show that all possible entropy functions of $\mathcal{G}$ form a convex set, and hence, there exists a continuum of possible entropy functions for $\mathcal{G}$ ranging from the required entropy supply $\mathcal{S}_r(E)$ to the available entropy $\mathcal{S}_a(E)$.

Proposition 3.3 *Consider the large-scale dynamical system $\mathcal{G}$ with power balance equation (3.5), and assume that Axioms i) and ii) hold. Then*

$$\mathcal{S}(E) \triangleq \alpha \mathcal{S}_r(E) + (1 - \alpha)\mathcal{S}_a(E), \quad \alpha \in [0, 1], \qquad (3.52)$$

is an entropy function for $\mathcal{G}$.

Proof. The result is a direct consequence of the reachability of $\mathcal{G}$ along with inequality (3.29) by noting that if $\mathcal{S}_r(E)$ and $\mathcal{S}_a(E)$ satisfy (3.29), then $\mathcal{S}(E)$ satisfies (3.29). $\qquad\qquad\square$

It follows from Proposition 3.3 that Definition 3.2 does not provide enough information to define the entropy uniquely for nonequilibrium thermodynamic systems with power balance equation (3.5). This difficulty has long been pointed out in [73]. Two particular entropy functions for $\mathcal{G}$ can be computed a priori via the variational problems given by (3.30) and (3.31). For equilibrium thermodynamics, however, uniqueness is not an issue, as shown in the next proposition.

Proposition 3.4 *Consider the large-scale dynamical system $\mathcal{G}$ with power balance equation (3.5), and assume that Axioms i) and ii) hold. Then at every equilibrium state $E = E_e$ of the isolated system $\mathcal{G}$, the entropy $\mathcal{S}(E)$, $E \in \overline{\mathbb{R}}_+^q$, of $\mathcal{G}$ is unique (modulo a constant of integration) and is given by*

$$\mathcal{S}(E) - \mathcal{S}(0) = \mathcal{S}_a(E) = \mathcal{S}_r(E) = \mathbf{e}^T \mathbf{log}_e(c\mathbf{e} + E) - q \log_e c, \quad (3.53)$$

where $E = E_e$ and $\mathbf{log}_e(c\mathbf{e} + E)$ denotes the vector natural logarithm given by $[\log_e(c + E_1), ..., \log_e(c + E_q)]^T$.

Proof. It follows from Axiom i) that for an equilibrium process $\phi_{ij}(E(t)) \equiv 0$, $i \neq j$, $i, j = 1, ..., q$. Consider the entropy function $\mathcal{S}_a(\cdot)$ given by (3.30), and let $E_0 = E_e$ for some equilibrium state E_e. Then it follows from (3.5) that

$$\mathcal{S}_a(E_0) = - \sup_{S(\cdot) \in \mathcal{U}_c, T \geq t_0} \int_{t_0}^{T} \sum_{i=1}^{q} \frac{\dot{E}_i(t) - \sum_{j=1, j \neq i}^{q} \phi_{ij}(E(t))}{c + E_i(t)} dt$$

$$= - \sup_{S(\cdot) \in \mathcal{U}_c, T \geq t_0} \left[\sum_{i=1}^{q} \log_e \left(\frac{c}{c + E_{i0}} \right) \right.$$

$$
\left. - \int_{t_0}^{T} \sum_{i=1}^{q} \sum_{j=1,j\neq i}^{q} \frac{\phi_{ij}(E(t))}{c + E_i(t)} \mathrm{d}t \right]
$$

$$
= \sum_{i=1}^{q} \log_e \left(\frac{c + E_{i0}}{c} \right) + \inf_{S(\cdot)\in\mathcal{U}_c, T\geq t_0} \int_{t_0}^{T} \sum_{i=1}^{q} \sum_{j=1,j\neq i}^{q} \frac{\phi_{ij}(E(t))}{c + E_i(t)} \mathrm{d}t
$$

$$
= \sum_{i=1}^{q} \log_e \left(\frac{c + E_{i0}}{c} \right)
$$

$$
+ \inf_{S(\cdot)\in\mathcal{U}_c, T\geq t_0} \int_{t_0}^{T} \sum_{i=1}^{q} \sum_{j=i+1}^{q} \left[\frac{\phi_{ij}(E(t))}{c + E_i(t)} - \frac{\phi_{ij}(E(t))}{c + E_j(t)} \right] \mathrm{d}t
$$

$$
= \sum_{i=1}^{q} \log_e \left(\frac{c + E_{i0}}{c} \right)
$$

$$
+ \inf_{S(\cdot)\in\mathcal{U}_c, T\geq t_0} \int_{t_0}^{T} \sum_{i=1}^{q} \sum_{j=i+1}^{q} \frac{\phi_{ij}(E(t))(E_j(t) - E_i(t))}{(c + E_i(t))(c + E_j(t))} \mathrm{d}t.
$$

$$(3.54)$$

Since the solution $E(t)$, $t \geq t_0$, to (3.5) is nonnegative for all nonnegative initial conditions, it follows from Axiom $ii)$ that the infimum in (3.54) is taken over the set of nonnegative values. However, the zero value of the infimum is achieved on an equilibrium process for which $\phi_{ij}(E(t)) \equiv 0$, $i \neq j$, $i,j = 1,...,q$. Thus,

$$
\mathcal{S}_{\mathrm{a}}(E_0) = \mathbf{e}^{\mathrm{T}}\mathbf{log}_e(ce + E_0) - q\log_e c, \quad E_0 = E_{\mathrm{e}}. \qquad (3.55)
$$

Similarly, consider the entropy function $\mathcal{S}_{\mathrm{r}}(\cdot)$ given by (3.31). Then, it follows from (3.5) that, for $E_0 = E_{\mathrm{e}}$,

$$
\mathcal{S}_{\mathrm{r}}(E_0) = \sup_{S(\cdot)\in\mathcal{U}_{\mathrm{r}}, T\geq -t_0} \int_{-T}^{t_0} \sum_{i=1}^{q} \frac{\dot{E}_i(t) - \sum_{j=1,j\neq i}^{q} \phi_{ij}(E(t))}{c + E_i(t)} \mathrm{d}t
$$

$$
= \sup_{S(\cdot)\in\mathcal{U}_{\mathrm{r}}, T\geq -t_0} \left[\sum_{i=1}^{q} \log_e \left(\frac{c + E_{i0}}{c} \right) \right.
$$

$$
\left. - \int_{-T}^{t_0} \sum_{i=1}^{q} \sum_{j=1,j\neq i}^{q} \frac{\phi_{ij}(E(t))}{c + E_i(t)} \mathrm{d}t \right]
$$

$$
= \sum_{i=1}^{q} \log_e \left(\frac{c + E_{i0}}{c} \right)
$$

$$- \inf_{S(\cdot) \in \mathcal{U}_r, T \geq -t_0} \int_{-T}^{t_0} \sum_{i=1}^{q} \sum_{j=1, j \neq i}^{q} \frac{\phi_{ij}(E(t))}{c + E_i(t)} dt$$

$$= \sum_{i=1}^{q} \log_e \left(\frac{c + E_{i0}}{c} \right)$$

$$- \inf_{S(\cdot) \in \mathcal{U}_r, T \geq -t_0} \int_{-T}^{t_0} \sum_{i=1}^{q} \sum_{j=i+1}^{q} \left[\frac{\phi_{ij}(E(t))}{c + E_i(t)} - \frac{\phi_{ij}(E(t))}{c + E_j(t)} \right] dt$$

$$= \sum_{i=1}^{q} \log_e \left(\frac{c + E_{i0}}{c} \right)$$

$$- \inf_{S(\cdot) \in \mathcal{U}_r, T \geq -t_0} \int_{-T}^{t_0} \sum_{i=1}^{q} \sum_{j=i+1}^{q} \frac{\phi_{ij}(E(t))(E_j(t) - E_i(t))}{(c + E_i(t))(c + E_j(t))} dt.$$

$$(3.56)$$

Now, it follows from Axioms i) and ii) that the zero value of the infimum in (3.56) is achieved on an equilibrium process and thus

$$S_r(E_0) = \mathbf{e}^T \log_e(c\mathbf{e} + E_0) - q \log_e c, \quad E_0 = E_e. \quad (3.57)$$

Finally, it follows from (3.32) that (3.53) holds. $\qquad \square$

The next proposition shows that if (3.29) holds as an equality for some transformation starting and ending at an equilibrium point of the isolated dynamical system $\mathcal{G}$, then this transformation must lie on the equilibrium manifold $\mathcal{M}_e$.

Proposition 3.5 *Consider the large-scale dynamical system $\mathcal{G}$ with power balance equation (3.5), and assume that Axioms i) and ii) hold. Let $S(\cdot)$ denote an entropy of $\mathcal{G}$, and let $E : [t_0, t_1] \to \overline{\mathbb{R}}_+^q$ denote the solution to (3.5) with $E(t_0) = \alpha_0 \mathbf{e}$ and $E(t_1) = \alpha_1 \mathbf{e}$, where $\alpha_0, \alpha_1 \geq 0$. Then*

$$S(E(t_1)) = S(E(t_0)) + \int_{t_0}^{t_1} \sum_{i=1}^{q} \frac{S_i(t) - \sigma_{ii}(E(t))}{c + E_i(t)} dt \quad (3.58)$$

if and only if there exists a continuous function $\alpha : [t_0, t_1] \to \overline{\mathbb{R}}_+$ such that $\alpha(t_0) = \alpha_0$, $\alpha(t_1) = \alpha_1$, and $E(t) = \alpha(t)\mathbf{e}$, $t \in [t_0, t_1]$.

Proof. Since $E(t_0)$ and $E(t_1)$ are equilibrium states of the isolated dynamical system $\mathcal{G}$, it follows from Proposition 3.4 that

$$S(E(t_1)) - S(E(t_0)) = q \log_e(c + \alpha_1) - q \log_e(c + \alpha_0). \quad (3.59)$$

Furthermore, it follows from (3.5) that

$$\int_{t_0}^{t_1} \sum_{i=1}^{q} \frac{S_i(t) - \sigma_{ii}(E(t))}{c + E_i(t)} dt$$

$$= \int_{t_0}^{t_1} \sum_{i=1}^{q} \frac{\dot{E}_i(t) - \sum_{j=1, j \neq i}^{q} \phi_{ij}(E(t))}{c + E_i(t)} dt$$

$$= q \log_e \left(\frac{c + \alpha_1}{c + \alpha_0} \right)$$

$$- \int_{t_0}^{t_1} \sum_{i=1}^{q} \sum_{j=i+1}^{q} \frac{\phi_{ij}(E(t))(E_j(t) - E_i(t))}{(c + E_i(t))(c + E_j(t))} dt. \quad (3.60)$$

Now, it follows from Axioms i) and ii) that (3.58) holds if and only if $E_i(t) = E_j(t)$, $t \in [t_0, t_1]$, $i \neq j$, $i, j = 1, ..., q$, or, equivalently, there exists a continuous function $\alpha : [t_0, t_1] \to \overline{\mathbb{R}}_+$ such that $E(t) = \alpha(t)\mathbf{e}$, $t \in [t_0, t_1]$, $\alpha(t_0) = \alpha_0$, and $\alpha(t_1) = \alpha_1$. $\qquad\square$

Even though it follows from Proposition 3.3 that Definition 3.2 does not provide a unique *continuous* entropy function for nonequilibrium systems, the next theorem gives a *unique, continuously differentiable* entropy function for $\mathcal{G}$ for equilibrium and nonequilibrium processes. This result answers the long-standing question of how the entropy of a nonequilibrium state of a dynamical process should be defined [63, 73], and establishes its global existence and uniqueness.

Theorem 3.4 *Consider the large-scale dynamical system $\mathcal{G}$ with power balance equation (3.5), and assume that Axioms i) and ii) hold. Then the function $\mathcal{S} : \overline{\mathbb{R}}_+^q \to \overline{\mathbb{R}}_+^q$ given by*

$$\mathcal{S}(E) = \mathbf{e}^{\mathrm{T}} \log_e(c\mathbf{e} + E) - q \log_e c, \quad E \in \overline{\mathbb{R}}_+^q, \quad (3.61)$$

where $c > 0$, is a unique (modulo a constant of integration), continuously differentiable entropy function of $\mathcal{G}$. Furthermore, for $E(t) \notin \mathcal{M}_e$, $t \geq t_0$, where $E(t)$, $t \geq t_0$, denotes the solution to (3.5) and $\mathcal{M}_e = \{ E \in \overline{\mathbb{R}}_+^q : E = \alpha\mathbf{e}, \alpha \geq 0 \}$, (3.61) satisfies

$$\mathcal{S}(E(t_2)) > \mathcal{S}(E(t_1)) + \int_{t_1}^{t_2} \sum_{i=1}^{q} \frac{S_i(t) - \sigma_{ii}(E(t))}{c + E_i(t)} dt \quad (3.62)$$

for any $t_2 \geq t_1 \geq t_0$ and $S(\cdot) \in \mathcal{U}$.

Proof. Since, by Proposition 3.1, $E(t) \geq\geq 0$, $t \geq t_0$, and $\phi_{ij}(E) = -\phi_{ji}(E)$, $E \in \overline{\mathbb{R}}_+^q$, $i \neq j$, $i, j = 1, ..., q$, it follows that

$$
\dot{\mathcal{S}}(E(t)) = \sum_{i=1}^{q} \frac{\dot{E}_i(t)}{c + E_i(t)}
$$

$$
= \sum_{i=1}^{q} \left[\frac{S_i(t) - \sigma_{ii}(E(t))}{c + E_i(t)} + \sum_{j=1, j \neq i}^{q} \frac{\phi_{ij}(E(t))}{c + E_i(t)} \right]
$$

$$
= \sum_{i=1}^{q} \left[\frac{S_i(t) - \sigma_{ii}(E(t))}{c + E_i(t)} + \sum_{j=i+1}^{q} \left(\frac{\phi_{ij}(E(t))}{c + E_i(t)} - \frac{\phi_{ij}(E(t))}{c + E_j(t)} \right) \right]
$$

$$
= \sum_{i=1}^{q} \frac{S_i(t) - \sigma_{ii}(E(t))}{c + E_i(t)} + \sum_{i=1}^{q} \sum_{j=i+1}^{q} \frac{\phi_{ij}(E(t))(E_j(t) - E_i(t))}{(c + E_i(t))(c + E_j(t))}
$$

$$
\geq \sum_{i=1}^{q} \frac{S_i(t) - \sigma_{ii}(E(t))}{c + E_i(t)}, \quad t \geq t_0. \tag{3.63}
$$

Now, integrating (3.63) over $[t_1, t_2]$ yields (3.29). Furthermore, in the case where $E(t) \notin \mathcal{M}_e$, $t \geq t_0$, it follows from Axiom i), Axiom ii), and (3.63) that (3.62) holds.

To show that (3.61) is a unique, continuously differentiable entropy function of $\mathcal{G}$, let $\mathcal{S}(E)$ be a continuously differentiable entropy function of $\mathcal{G}$ so that $\mathcal{S}(E)$ satisfies (3.29) or, equivalently,

$$
\dot{\mathcal{S}}(E(t)) \geq \mu^{\mathrm{T}}(E(t))[S(t) - d(E(t))], \quad t \geq t_0, \tag{3.64}
$$

where $\mu^{\mathrm{T}}(E) = [\frac{1}{c+E_1}, ..., \frac{1}{c+E_q}]$, $E \in \overline{\mathbb{R}}_+^q$, $E(t)$, $t \geq t_0$, denotes the solution to the power balance equation (3.5), and $\dot{\mathcal{S}}(E(t))$ denotes the time derivative of $\mathcal{S}(E)$ along the solution $E(t)$, $t \geq t_0$. Hence, it follows from (3.64) that

$$
\mathcal{S}'(E)[w(E) - d(E) + S] \geq \mu^{\mathrm{T}}(E)[S - d(E)], \quad E \in \overline{\mathbb{R}}_+^q, \quad S \in \mathbb{R}^q, \tag{3.65}
$$

which implies that there exist continuous functions $\ell : \overline{\mathbb{R}}_+^q \to \mathbb{R}^p$ and $\mathcal{W} : \overline{\mathbb{R}}_+^q \to \mathbb{R}^{p \times q}$ such that

$$
\begin{aligned}
0 = &\mathcal{S}'(E)[w(E) - d(E) + S] - \mu^{\mathrm{T}}(E)[S - d(E)] \\
&- [\ell(E) + \mathcal{W}(E)S]^{\mathrm{T}}[\ell(E) + \mathcal{W}(E)S], \quad E \in \overline{\mathbb{R}}_+^q, \quad S \in \mathbb{R}^q.
\end{aligned} \tag{3.66}
$$

Now, equating coefficients of equal powers (of S), it follows that $\mathcal{W}(E)$ $\equiv 0$, $\mathcal{S}'(E) = \mu^{\mathrm{T}}(E)$, $E \in \overline{\mathbb{R}}_+^q$, and

$$0 = \mathcal{S}'(E)w(E) - \ell^{\mathrm{T}}(E)\ell(E), \quad E \in \overline{\mathbb{R}}_+^q. \tag{3.67}$$

Hence, $\mathcal{S}(E) = \mathbf{e}^{\mathrm{T}}\log_e(c\mathbf{e} + E) - q\log_e c$, $E \in \overline{\mathbb{R}}_+^q$, and

$$0 = \mu^{\mathrm{T}}(E)w(E) - \ell^{\mathrm{T}}(E)\ell(E), \quad E \in \overline{\mathbb{R}}_+^q. \tag{3.68}$$

Thus, (3.61) is a unique, continuously differentiable entropy function for $\mathcal{G}$. $\square$

Note that it follows from Axiom i), Axiom ii), and the last equality in (3.63) that the entropy function given by (3.61) satisfies (3.29) as an equality for an equilibrium process and as a strict inequality for a nonequilibrium process. Hence, it follows from Theorem 2.15 that the isolated (i.e., $S(t) \equiv 0$ and $d(E) \equiv 0$) large-scale dynamical system $\mathcal{G}$ does not exhibit Poincaré recurrence in $\overline{\mathbb{R}}_+^q \setminus \mathcal{M}_e$. Furthermore, for any entropy function of $\mathcal{G}$, it follows from Proposition 3.5 that if (3.29) holds as an equality for some transformation starting and ending at equilibrium points of the isolated system $\mathcal{G}$, then this transformation must lie on the equilibrium manifold $\mathcal{M}_e$. However, (3.29) may hold as an equality for nonequilibrium processes starting and ending at nonequilibrium states. The entropy expression given by (3.61) is identical in form to the Boltzmann entropy for statistical thermodynamics. Due to the fact that the entropy given by (3.61) is indeterminate to the extent of an additive constant, we can place the constant of integration $q\log_e c$ to zero by taking $c = 1$. Since $\mathcal{S}(E)$ given by (3.61) achieves a maximum when all the subsystem energies E_i, $i = 1, ..., q$, are equal, the entropy of $\mathcal{G}$ can be thought of as a measure of the tendency of a system to lose the ability to do useful work, lose order, and settle to a more homogenous state.

Recalling that $\mathrm{d}Q_i(t) = [S_i(t) - \sigma_{ii}(E(t))]\mathrm{d}t$, $i = 1, ..., q$, is the infinitesimal amount of the net heat received or dissipated by the ith subsystem of $\mathcal{G}$ over the infinitesimal time interval $\mathrm{d}t$, it follows from (3.29) that

$$\mathrm{d}\mathcal{S}(E(t)) \geq \sum_{i=1}^q \frac{\mathrm{d}Q_i(t)}{c + E_i(t)}, \quad t \geq t_0. \tag{3.69}$$

Inequality (3.69) is analogous to the classical thermodynamic inequality for the variation of entropy during an infinitesimal irreversible transformation with the shifted subsystem energies $c + E_i$ playing the role of the ith subsystem thermodynamic (absolute) temperatures. Specifically, note that since $\frac{d\mathcal{S}_i}{dE_i} = \frac{1}{c+E_i}$, where $\mathcal{S}_i = \log_e(c+E_i) - \log_e c$ denotes the unique continuously differentiable ith subsystem entropy, it follows that $\frac{d\mathcal{S}_i}{dE_i}$, $i = 1, ..., q$, defines the reciprocal of the subsystem thermodynamic temperatures. That is,

$$\frac{1}{T_i} \triangleq \frac{d\mathcal{S}_i}{dE_i} \tag{3.70}$$

and $T_i > 0$, $i = 1, ..., q$. Hence, in our formulation, temperature is a function derived from entropy and does not involve the primitive subjective notions of hotness and coldness.

It is important to note that in this chapter we view subsystem temperatures to be synonymous with subsystem energies. Even though this does not limit the generality of our theory from a mathematical perspective, it can be physically limiting since it does not allow for the consideration of two subsystems of $\mathcal{G}$ having the same stored energy with one of the subsystems being at a higher temperature (i.e., *hotter*) than the other. This, however, can be easily addressed by assigning different specific heats (i.e., thermal capacities) for each of the compartments of the large-scale system $\mathcal{G}$ as shown in Chapter 4.

Finally, using the system entropy function given by (3.61), we show that our large-scale dynamical system $\mathcal{G}$ with power balance equation (3.5) is state irreversible for every nontrivial (nonequilibrium) trajectory of $\mathcal{G}$. For this result, let $\mathcal{W}_{[t_0, t_1]}$ denote the set of all possible energy trajectories of $\mathcal{G}$ over the time interval $[t_0, t_1]$ given by

$$\mathcal{W}_{[t_0, t_1]} \triangleq \{s^E : [t_0, t_1] \times \mathcal{U} \to \overline{\mathbb{R}}_+^q : s^E(\cdot, S(\cdot)) \text{ satisfies } (3.5)\}, \tag{3.71}$$

and let $\mathcal{M}_e \subset \overline{\mathbb{R}}_+^q$ denote the set of equilibria of the isolated system $\mathcal{G}$ given by $\mathcal{M}_e = \{E \in \overline{\mathbb{R}}_+^q : \alpha\mathbf{e}, \alpha \geq 0\}$.

Theorem 3.5 *Consider the large-scale dynamical system $\mathcal{G}$ with power balance equation (3.5), and assume Axioms i) and ii) hold. Furthermore, let $s^E(\cdot, S(\cdot)) \in \mathcal{W}_{[t_0, t_1]}$, where $S(\cdot) \in \mathcal{U}$. Then $s^E(\cdot, S(\cdot))$ is an I_q-reversible trajectory of $\mathcal{G}$ if and only if $s^E(t, S(t)) \in \mathcal{M}_e$, $t \in [t_0, t_1]$.*

Proof. First, note that it follows from Theorem 3.4 that if $E(t) \notin \mathcal{M}_e$, $t \geq t_0$, then there exists an entropy function $\mathcal{S}(E)$, $E \in \overline{\mathbb{R}}_+^q$, for $\mathcal{G}$ such that (3.62) holds. Now, sufficiency follows as a direct consequence of Theorem 2.7 with $R = I_q$, $V(z) = \mathcal{S}(E)$, and $r(u, y) = r(S, d(E)) = \sum_{i=1}^q \frac{S_i - \sigma_{ii}(E)}{c+E_i}$. To show necessity, assume that $s^E(t, S(t)) \in \mathcal{M}_e$, $t \in [t_0, t_1]$. In this case, it can be shown that $S(t) = d(E(t)) + u(t)$, $t \geq t_0$, where $u(\cdot) \in \mathcal{U}$ is such that $u_i(t) \equiv u_j(t)$, $i \neq j$, $i, j = 1, ..., q$. Now, with $S^-(t) = d(E(t)) + u^-(t)$, $t \geq t_0$, where $u^-(t) = -u(t_1 + t_0 - t)$, $t \in [t_0, t_1]$, it follows that $s^E(t, S(t))$ is an I_q-reversible trajectory of $\mathcal{G}$. $\square$

Theorem 3.5 establishes an equivalence between (non)equilibrium and state (ir)reversible thermodynamic systems. Furthermore, Theorem 3.5 shows that for every $E_0 \notin \mathcal{M}_e$, the large-scale dynamical system $\mathcal{G}$ is state irreversible. In addition, since state irrecoverability implies state irreversibility and, by Theorem 3.5, state irreversibility is equivalent to $E(t) \notin \mathcal{M}_e$, $t \geq t_0$, it follows from Theorem 2.8 that state (ir)reversibility and state (ir)recoverability are equivalent for our thermodynamically consistent large-scale dynamical system $\mathcal{G}$. Hence, in the remainder of the monograph we use the notions of (non)equilibrium, state (ir)reversible, and state (ir)recoverable dynamical processes interchangeably.

3.4 Ectropy

In this section, we introduce a *new* and dual notion to entropy, namely, ectropy, describing the status quo of the large-scale dynamical system $\mathcal{G}$. First, however, we present a dual inequality to inequality (3.26) that holds for our thermodynamically consistent energy flow model over cyclic motions.

Proposition 3.6 *Consider the large-scale dynamical system $\mathcal{G}$ with power balance equation (3.5), and assume that Axioms i) and ii) hold. Then for all $E_0 \in \overline{\mathbb{R}}_+^q$, $t_f \geq t_0$, and $S(\cdot) \in \mathcal{U}$ such that $E(t_f) = E(t_0) = E_0$,*

$$\int_{t_0}^{t_f} \sum_{i=1}^{q} E_i(t)[S_i(t) - \sigma_{ii}(E(t))]dt = \oint \sum_{i=1}^{q} E_i(t)dQ_i(t) \geq 0, \quad (3.72)$$

where $E(t)$, $t \geq t_0$, is the solution to (3.5) with initial condition $E(t_0) = E_0$. Furthermore,

$$\oint \sum_{i=1}^{q} E_i(t)dQ_i(t) = 0 \qquad (3.73)$$

if and only if there exists a continuous function $\alpha : [t_0, t_f] \to \overline{\mathbb{R}}_+$ such that $E(t) = \alpha(t)e$, $t \in [t_0, t_f]$.

Proof. Since, by Proposition 3.1, $E(t) \geq\geq 0$, $t \geq t_0$, and $\phi_{ij}(E) = -\phi_{ji}(E)$, $E \in \overline{\mathbb{R}}_+^q$, $i \neq j$, $i, j = 1, ..., q$, it follows from (3.5) and Axiom ii) that

$$
\oint \sum_{i=1}^{q} E_i(t) dQ_i(t) = \int_{t_0}^{t_f} \sum_{i=1}^{q} E_i(t)[\dot{E}_i(t) - \sum_{j=1, j \neq i}^{q} \phi_{ij}(E(t))] dt
$$
$$
= \tfrac{1}{2} E^{\mathrm{T}}(t_f) E(t_f) - \tfrac{1}{2} E^{\mathrm{T}}(t_0) E(t_0)
$$
$$
- \int_{t_0}^{t_f} \sum_{i=1}^{q} \sum_{j=1, j \neq i}^{q} E_i(t) \phi_{ij}(E(t)) dt
$$
$$
= - \int_{t_0}^{t_f} \sum_{i=1}^{q} \sum_{j=i+1}^{q} \phi_{ij}(E(t))[E_i(t) - E_j(t)] dt
$$
$$
\geq 0, \tag{3.74}
$$

which proves (3.72).

To show (3.73), note that it follows from (3.74), Axiom i), and Axiom ii) that (3.73) holds if and only if $E_i(t) = E_j(t)$, $i \neq j$, $i, j = 1, ..., q$, or, equivalently, there exists a continuous function $\alpha : [t_0, t_f] \to \overline{\mathbb{R}}_+$ such that $E(t) = \alpha(t) e$, $t \in [t_0, t_f]$. $\square$

Inequality (3.72) is an anti–Clausius inequality and restricts the manner in which the system absorbs (scaled) heat over cyclic motions. Note that inequality (3.72) is satisfied as an equality for an equilibrium process and as a strict inequality for a nonequilibrium process. Next, we present the definition of ectropy for the large-scale dynamical system $\mathcal{G}$.

Definition 3.3 *For the large-scale dynamical system $\mathcal{G}$ with power balance equation (3.5), a function $\mathcal{E} : \overline{\mathbb{R}}_+^q \to \mathbb{R}$ satisfying*

$$
\mathcal{E}(E(t_2)) \leq \mathcal{E}(E(t_1)) + \int_{t_1}^{t_2} \sum_{i=1}^{q} E_i(t)[S_i(t) - \sigma_{ii}(E(t))] dt \tag{3.75}
$$

for any $t_2 \geq t_1 \geq t_0$ and $S(\cdot) \in \mathcal{U}$ is called the ectropy *function of $\mathcal{G}$.*

For the next result, define the *available ectropy* of the large-scale dynamical system $\mathcal{G}$ by

$$
\mathcal{E}_a(E_0) \triangleq - \inf_{S(\cdot) \in \mathcal{U}_c, T \geq t_0} \int_{t_0}^{T} \sum_{i=1}^{q} E_i(t)[S_i(t) - \sigma_{ii}(E(t))] dt, \tag{3.76}
$$

where $E(t_0) = E_0 \in \overline{\mathbb{R}}_+^q$ and $E(T) = 0$, and define the *required ectropy supply* of the large-scale dynamical system $\mathcal{G}$ by

$$\mathcal{E}_r(E_0) \triangleq \inf_{S(\cdot) \in \mathcal{U}_r, \, T \geq -t_0} \int_{-T}^{t_0} \sum_{i=1}^{q} E_i(t)[S_i(t) - \sigma_{ii}(E(t))]dt, \quad (3.77)$$

where $E(-T) = 0$ and $E(t_0) = E_0 \in \overline{\mathbb{R}}_+^q$. Note that the available ectropy $\mathcal{E}_a(E_0)$ is the maximum amount of scaled heat (ectropy) that can be extracted from the large-scale dynamical system $\mathcal{G}$ in order to transfer it from an initial state $E(t_0) = E_0$ to $E(T) = 0$. Alternatively, the required ectropy supply $\mathcal{E}_r(E_0)$ is the minimum amount of scaled heat (ectropy) that can be delivered to $\mathcal{G}$ to transfer it from an initial state $E(-T) = 0$ to a given state $E(t_0) = E_0$.

Theorem 3.6 *Consider the large-scale dynamical system $\mathcal{G}$ with power balance equation (3.5), and assume that Axiom ii) holds. Then there exists an ectropy function for $\mathcal{G}$. Moreover, $\mathcal{E}_a(E)$, $E \in \overline{\mathbb{R}}_+^q$, and $\mathcal{E}_r(E)$, $E \in \overline{\mathbb{R}}_+^q$, are possible ectropy functions for $\mathcal{G}$ with $\mathcal{E}_a(0) = \mathcal{E}_r(0) = 0$. Finally, all ectropy functions $\mathcal{E}(E)$, $E \in \overline{\mathbb{R}}_+^q$, for $\mathcal{G}$ satisfy*

$$\mathcal{E}_a(E) \leq \mathcal{E}(E) - \mathcal{E}(0) \leq \mathcal{E}_r(E), \quad E \in \overline{\mathbb{R}}_+^q. \quad (3.78)$$

Proof. Since, by Lemma 3.1, $\mathcal{G}$ is controllable to and reachable from the origin in $\overline{\mathbb{R}}_+^q$, it follows from (3.76) and (3.77) that $\mathcal{E}_a(E_0) > -\infty$, $E_0 \in \overline{\mathbb{R}}_+^q$, and $\mathcal{E}_r(E_0) < \infty$, $E_0 \in \overline{\mathbb{R}}_+^q$, respectively. Next, let $E_0 \in \overline{\mathbb{R}}_+^q$, and let $S(\cdot) \in \mathcal{U}$ be such that $E(t_i) = E(t_f) = 0$ and $E(t_0) = E_0$, where $t_i < t_0 < t_f$. In this case, it follows from (3.72) that

$$\int_{t_i}^{t_f} \sum_{i=1}^{q} E_i(t)[S_i(t) - \sigma_{ii}(E(t))]dt \geq 0 \quad (3.79)$$

or, equivalently,

$$\int_{t_i}^{t_0} \sum_{i=1}^{q} E_i(t)[S_i(t) - \sigma_{ii}(E(t))]dt$$

$$\geq -\int_{t_0}^{t_f} \sum_{i=1}^{q} E_i(t)[S_i(t) - \sigma_{ii}(E(t))]dt. \quad (3.80)$$

Now, taking the infimum on both sides of (3.80) over all $S(\cdot) \in \mathcal{U}_r$ and $t_i \leq t_0$ yields

$$\mathcal{E}_r(E_0) = \inf_{S(\cdot) \in \mathcal{U}_r, \, t_i \leq t_0} \int_{t_i}^{t_0} \sum_{i=1}^{q} E_i(t)[S_i(t) - \sigma_{ii}(E(t))]dt$$

$$\geq -\int_{t_0}^{t_f} \sum_{i=1}^{q} E_i(t)[S_i(t) - \sigma_{ii}(E(t))]dt. \tag{3.81}$$

Next, taking the supremum on both sides of (3.81) over all $S(\cdot) \in \mathcal{U}_c$ and $t_f \geq t_0$, we obtain $\mathcal{E}_r(E_0) \geq \mathcal{E}_a(E_0)$, $E_0 \in \overline{\mathbb{R}}_+^q$, which implies that $-\infty < \mathcal{E}_a(E_0) \leq \mathcal{E}_r(E_0) < \infty$, $E_0 \in \overline{\mathbb{R}}_+^q$. Hence, the functions $\mathcal{E}_a(\cdot)$ and $\mathcal{E}_r(\cdot)$ are well defined.

Next, it follows from the definition of $\mathcal{E}_a(\cdot)$ that, for any $T \geq t_1$ and $S(\cdot) \in \mathcal{U}_c$ such that $E(t_1) \in \overline{\mathbb{R}}_+^q$ and $E(T) = 0$,

$$-\mathcal{E}_a(E(t_1)) \leq \int_{t_1}^{t_2} \sum_{i=1}^{q} E_i(t)[S_i(t) - \sigma_{ii}(E(t))]dt$$

$$+ \int_{t_2}^{T} \sum_{i=1}^{q} E_i(t)[S_i(t) - \sigma_{ii}(E(t))]dt, \quad t_1 \leq t_2 \leq T,$$

$$\tag{3.82}$$

and hence,

$$-\mathcal{E}_a(E(t_1)) \leq \int_{t_1}^{t_2} \sum_{i=1}^{q} E_i(t)[S_i(t) - \sigma_{ii}(E(t))]dt$$

$$+ \inf_{S(\cdot) \in \mathcal{U}_c, T \geq t_2} \int_{t_2}^{T} \sum_{i=1}^{q} E_i(t)[S_i(t) - \sigma_{ii}(E(t))]dt$$

$$= \int_{t_1}^{t_2} \sum_{i=1}^{q} E_i(t)[S_i(t) - \sigma_{ii}(E(t))]dt - \mathcal{E}_a(E(t_2)), \tag{3.83}$$

which implies that $\mathcal{E}_a(E)$, $E \in \overline{\mathbb{R}}_+^q$, satisfies (3.75). Thus, $\mathcal{E}_a(E)$, $E \in \overline{\mathbb{R}}_+^q$, is a possible ectropy function for $\mathcal{G}$. Note that with $E(t_0) = E(T) = 0$ it follows from (3.72) that the infimum in (3.76) is taken over the set of nonnegative values with one of the values being zero for $S(t) \equiv 0$. Thus, $\mathcal{E}_a(0) = 0$.

Similarly, it follows from the definition of $\mathcal{E}_r(\cdot)$ that, for any $T \geq -t_2$ and $S(\cdot) \in \mathcal{U}_r$ such that $E(t_2) \in \overline{\mathbb{R}}_+^q$ and $E(-T) = 0$,

$$\mathcal{E}_r(E(t_2)) \leq \int_{-T}^{t_1} \sum_{i=1}^{q} E_i(t)[S_i(t) - \sigma_{ii}(E(t))]dt$$

$$+ \int_{t_1}^{t_2} \sum_{i=1}^{q} E_i(t)[S_i(t) - \sigma_{ii}(E(t))]dt, \quad -T \leq t_1 \leq t_2,$$

$$\tag{3.84}$$

and hence,

$$\mathcal{E}_r(E(t_2)) \leq \int_{t_1}^{t_2} \sum_{i=1}^{q} E_i(t)[S_i(t) - \sigma_{ii}(E(t))]\mathrm{d}t$$

$$+ \inf_{S(\cdot)\in\mathcal{U}_r, T\geq -t_1} \int_{-T}^{t_1} \sum_{i=1}^{q} E_i(t)[S_i(t) - \sigma_{ii}(E(t))]\mathrm{d}t$$

$$= \int_{t_1}^{t_2} \sum_{i=1}^{q} E_i(t)[S_i(t) - \sigma_{ii}(E(t))]\mathrm{d}t + \mathcal{E}_r(E(t_1)), \quad (3.85)$$

which implies that $\mathcal{E}_r(E)$, $E \in \overline{\mathbb{R}}_+^q$, satisfies (3.75). Thus, $\mathcal{E}_r(E)$, $E \in \overline{\mathbb{R}}_+^q$, is a possible ectropy function for $\mathcal{G}$. Note that with $E(t_0) = E(-T) = 0$ it follows from (3.72) that the infimum in (3.77) is taken over the set of nonnegative values with one of the values being zero for $S(t) \equiv 0$. Thus, $\mathcal{E}_r(0) = 0$.

Next, suppose there exists an ectropy function $\mathcal{E} : \overline{\mathbb{R}}_+^q \to \mathbb{R}$ for $\mathcal{G}$, and let $E(t_2) = 0$ in (3.75). Then it follows from (3.75) that

$$\mathcal{E}(E(t_1)) - \mathcal{E}(0) \geq -\int_{t_1}^{t_2} \sum_{i=1}^{q} E_i(t)[S_i(t) - \sigma_{ii}(E(t))]\mathrm{d}t \quad (3.86)$$

for all $t_2 \geq t_1$ and $S(\cdot) \in \mathcal{U}_c$, which implies that

$$\mathcal{E}(E(t_1)) - \mathcal{E}(0) \geq \sup_{S(\cdot)\in\mathcal{U}_c, t_2\geq t_1} \left[-\int_{t_1}^{t_2} \sum_{i=1}^{q} E_i(t)[S_i(t) - \sigma_{ii}(E(t))]\mathrm{d}t \right]$$

$$= -\inf_{S(\cdot)\in\mathcal{U}_c, t_2\geq t_1} \int_{t_1}^{t_2} \sum_{i=1}^{q} E_i(t)[S_i(t) - \sigma_{ii}(E(t))]\mathrm{d}t$$

$$= \mathcal{E}_a(E(t_1)). \quad (3.87)$$

Since $E(t_1)$ is arbitrary, it follows that $\mathcal{E}(E) - \mathcal{E}(0) \geq \mathcal{E}_a(E)$, $E \in \overline{\mathbb{R}}_+^q$. Alternatively, let $E(t_1) = 0$ in (3.75). Then it follows from (3.75) that

$$\mathcal{E}(E(t_2)) - \mathcal{E}(0) \leq \int_{t_1}^{t_2} \sum_{i=1}^{q} E_i(t)[S_i(t) - \sigma_{ii}(E(t))]\mathrm{d}t \quad (3.88)$$

for all $t_1 \leq t_2$ and $S(\cdot) \in \mathcal{U}_r$. Hence,

$$\mathcal{E}(E(t_2)) - \mathcal{E}(0) \leq \inf_{S(\cdot)\in\mathcal{U}_r, t_1\leq t_2} \int_{t_1}^{t_2} \sum_{i=1}^{q} E_i(t)[S_i(t) - \sigma_{ii}(E(t))]\mathrm{d}t$$

$$= \mathcal{E}_r(E(t_2)), \quad (3.89)$$

which, since $E(t_2)$ is arbitrary, implies that $\mathcal{E}_r(E) \geq \mathcal{E}(E) - \mathcal{E}(0)$, $E \in \overline{\mathbb{R}}_+^q$. Thus, all ectropy functions for $\mathcal{G}$ satisfy (3.78). $\qquad\square$

The next result shows that all ectropy functions for $\mathcal{G}$ are continuous on $\overline{\mathbb{R}}_+^q$.

Theorem 3.7 *Consider the large-scale dynamical system $\mathcal{G}$ with power balance equation (3.5), and let $\mathcal{E} : \overline{\mathbb{R}}_+^q \to \mathbb{R}$ be an ectropy function of $\mathcal{G}$. Then $\mathcal{E}(\cdot)$ is continuous on $\overline{\mathbb{R}}_+^q$.*

Proof. The proof is identical to the proof of Theorem 3.3. $\qquad\square$

The next result is a direct consequence of Theorem 3.6.

Proposition 3.7 *Consider the large-scale dynamical system $\mathcal{G}$ with power balance equation (3.5), and assume that Axioms i) and ii) hold. Then*

$$\mathcal{E}(E) \triangleq \alpha \mathcal{E}_a(E) + (1 - \alpha)\mathcal{E}_r(E), \quad \alpha \in [0, 1], \tag{3.90}$$

is an ectropy function for $\mathcal{G}$.

As in the case of entropy, in the next proposition we show that any ectropy function for $\mathcal{G}$ has a unique form when evaluated on the set of equilibria $\mathcal{M}_e$ for the isolated large-scale dynamical system $\mathcal{G}$.

Proposition 3.8 *Consider the large-scale dynamical system $\mathcal{G}$ with power balance equation (3.5), and assume that Axioms i) and ii) hold. Then at every equilibrium state $E = E_e$ of the isolated system $\mathcal{G}$, the ectropy $\mathcal{E}(E)$, $E \in \overline{\mathbb{R}}_+^q$, of $\mathcal{G}$ is unique (modulo a constant of integration) and is given by*

$$\mathcal{E}(E) - \mathcal{E}(0) = \mathcal{E}_a(E) = \mathcal{E}_r(E) = \tfrac{1}{2}E^{\mathrm{T}}E, \quad E = E_e. \tag{3.91}$$

Proof. It follows from Axiom i) that for an equilibrium process $\phi_{ij}(E(t)) \equiv 0$, $i \neq j$, $i, j = 1, ..., q$. Consider the ectropy function $\mathcal{E}_a(\cdot)$ given by (3.76), and let $E_0 = E_e$ for some equilibrium state E_e. Then, it follows from (3.5) that

$$\mathcal{E}_a(E_0) = - \inf_{S(\cdot) \in \mathcal{U}_c, T \geq t_0} \int_{t_0}^T \sum_{i=1}^q E_i(t)[\dot{E}_i(t) - \sum_{j=1, j \neq i}^q \phi_{ij}(E(t))]\mathrm{d}t$$

$$= - \inf_{S(\cdot) \in \mathcal{U}_c, T \geq t_0} \left[- \sum_{i=1}^q \tfrac{1}{2}E_{i0}^2 \right.$$

$$-\int_{t_0}^{T}\sum_{i=1}^{q}\sum_{j=1,j\neq i}^{q}E_i(t)\phi_{ij}(E(t))\mathrm{d}t\Bigg]$$

$$=\sum_{i=1}^{q}\tfrac{1}{2}E_{i0}^{2}+\sup_{S(\cdot)\in\mathcal{U}_c,\,T\geq t_0}\int_{t_0}^{T}\sum_{i=1}^{q}\sum_{j=1,j\neq i}^{q}E_i(t)\phi_{ij}(E(t))\mathrm{d}t$$

$$=\tfrac{1}{2}E_0^{\mathrm{T}}E_0$$

$$+\sup_{S(\cdot)\in\mathcal{U}_c,\,T\geq t_0}\int_{t_0}^{T}\sum_{i=1}^{q}\sum_{j=i+1}^{q}[E_i(t)-E_j(t)]\phi_{ij}(E(t))\mathrm{d}t.$$

$$(3.92)$$

Since the solution $E(t)$, $t \geq t_0$, to (3.5) is nonnegative for all nonnegative initial conditions, it follows from Axiom ii) that the supremum in (3.92) is taken over the set of negative semi-definite values. However, the zero value of the supremum is achieved on an equilibrium process for which $\phi_{ij}(E(t)) \equiv 0$, $i \neq j$, $i, j = 1, ..., q$. Thus,

$$\mathcal{E}_\mathrm{a}(E_0) = \tfrac{1}{2}E_0^{\mathrm{T}}E_0, \quad E_0 = E_\mathrm{e}. \tag{3.93}$$

Similarly, it can be shown that $\mathcal{E}_\mathrm{r}(E) = \tfrac{1}{2}E^{\mathrm{T}}E$ for $E = E_\mathrm{e}$. Finally, it follows from (3.78) that (3.91) holds. $\qquad\square$

The next proposition shows that if (3.75) holds as an equality for some transformation starting and ending at equilibrium points of the isolated system $\mathcal{G}$, then this transformation must lie on the equilibrium manifold $\mathcal{M}_\mathrm{e}$.

Proposition 3.9 *Consider the large-scale dynamical system $\mathcal{G}$ with power balance equation (3.5), and assume that Axioms i) and ii) hold. Let $\mathcal{E}(\cdot)$ denote an ectropy of $\mathcal{G}$, and let $E : [t_0, t_1] \to \overline{\mathbb{R}}_+^q$ denote the solution to (3.5) with $E(t_0) = \alpha_0\mathbf{e}$ and $E(t_1) = \alpha_1\mathbf{e}$, where $\alpha_0, \alpha_1 \geq 0$. Then*

$$\mathcal{E}(E(t_1)) = \mathcal{E}(E(t_0)) + \int_{t_0}^{t_1}\sum_{i=1}^{q}E_i(t)[S_i(t) - \sigma_{ii}(E(t))]\mathrm{d}t \tag{3.94}$$

if and only if there exists a continuous function $\alpha : [t_0, t_1] \to \overline{\mathbb{R}}_+$ such that $\alpha(t_0) = \alpha_0$, $\alpha(t_1) = \alpha_1$, and $E(t) = \alpha(t)\mathbf{e}$, $t \in [t_0, t_1]$.

Proof. Since $E(t_0)$ and $E(t_1)$ are equilibrium states of the isolated dynamical system $\mathcal{G}$, it follows from Proposition 3.8 that

$$\mathcal{E}(E(t_1)) - \mathcal{E}(E(t_0)) = \tfrac{1}{2}E^{\mathrm{T}}(t_1)E(t_1) - \tfrac{1}{2}E^{\mathrm{T}}(t_0)E(t_0). \tag{3.95}$$

Furthermore, it follows from (3.5) that

$$\int_{t_0}^{t_1} \sum_{i=1}^{q} E_i(t)[S_i(t) - \sigma_{ii}(E(t))]dt$$

$$= \int_{t_0}^{t_1} \sum_{i=1}^{q} E_i(t)[\dot{E}_i(t) - \sum_{j=1, j\neq i}^{q} \phi_{ij}(E(t))]dt$$

$$= \tfrac{1}{2}q\alpha_1^2 - \tfrac{1}{2}q\alpha_0^2$$

$$- \int_{t_0}^{t_1} \sum_{i=1}^{q} \sum_{j=i+1}^{q} \phi_{ij}(E(t))[E_i(t) - E_j(t)]dt. \quad (3.96)$$

Now, it follows from Axioms i) and ii) that (3.94) holds if and only if $E_i(t) = E_j(t)$, $t \in [t_0, t_1]$, $i \neq j$, $i, j = 1, ..., q$, or, equivalently, there exists a continuous function $\alpha : [t_0, t_1] \to \overline{\mathbb{R}}_+$ such that $E(t) = \alpha(t)\mathbf{e}$, $t \in [t_0, t_1]$, $\alpha(t_0) = \alpha_0$, and $\alpha(t_1) = \alpha_1$. $\square$

The next theorem gives a unique, continuously differentiable ectropy function for $\mathcal{G}$ for equilibrium and nonequilibrium processes.

Theorem 3.8 *Consider the large-scale dynamical system $\mathcal{G}$ with power balance equation (3.5), and assume that Axioms i) and ii) hold. Then the function $\mathcal{E} : \overline{\mathbb{R}}_+^q \to \overline{\mathbb{R}}_+$ given by*

$$\mathcal{E}(E) = \tfrac{1}{2}E^{\mathrm{T}}E, \quad E \in \overline{\mathbb{R}}_+^q, \quad (3.97)$$

is a unique (modulo a constant of integration), continuously differentiable ectropy function of $\mathcal{G}$. Furthermore, for $E(t) \notin \mathcal{M}_e$, $t \geq t_0$, where $E(t)$, $t \geq t_0$, denotes the solution to (3.5) and $\mathcal{M}_e = \{E \in \overline{\mathbb{R}}_+^q : E = \alpha\mathbf{e}, \alpha \geq 0\}$, (3.97) satisfies

$$\mathcal{E}(E(t_2)) < \mathcal{E}(E(t_1)) + \int_{t_1}^{t_2} \sum_{i=1}^{q} E_i(t)[S_i(t) - \sigma_{ii}(E(t))]dt \quad (3.98)$$

for any $t_2 \geq t_1 \geq t_0$ and $S(\cdot) \in \mathcal{U}$.

Proof. Since, by Proposition 3.1, $E(t) \geq\geq 0$, $t \geq t_0$, and $\phi_{ij}(E) = -\phi_{ji}(E)$, $E \in \overline{\mathbb{R}}_+^q$, $i \neq j$, $i, j = 1, ..., q$, it follows that

$$\dot{\mathcal{E}}(E(t)) = \sum_{i=1}^{q} \dot{E}_i(t)E_i(t)$$

$$= \sum_{i=1}^{q} E_i(t)[S_i(t) - \sigma_{ii}(E(t))] + \sum_{i=1}^{q} \sum_{j=1, j\neq i}^{q} E_i(t)\phi_{ij}(E(t))$$

$$= \sum_{i=1}^{q} E_i(t)[S_i(t) - \sigma_{ii}(E(t))]$$

$$+ \sum_{i=1}^{q} \sum_{j=i+1}^{q} [E_i(t) - E_j(t)]\phi_{ij}(E(t))$$

$$\leq \sum_{i=1}^{q} E_i(t)[S_i(t) - \sigma_{ii}(E(t))], \quad t \geq t_0. \tag{3.99}$$

Now, integrating (3.99) over $[t_1, t_2]$ yields (3.75). Furthermore, in the case where $E(t) \notin \mathcal{M}_e$, $t \geq t_0$, it follows from Axiom i), Axiom ii), and (3.99) that (3.98) holds.

To show that (3.97) is a unique, continuously differentiable ectropy function of $\mathcal{G}$, let $\mathcal{E}(E)$ be a continuously differentiable ectropy function of $\mathcal{G}$ so that $\mathcal{E}(E)$ satisfies (3.75) or, equivalently,

$$\dot{\mathcal{E}}(E(t)) \leq E^{\mathrm{T}}(t)[S(t) - d(E(t))], \quad t \geq t_0, \tag{3.100}$$

where $E(t)$, $t \geq t_0$, denotes the solution to the power balance equation (3.5) and $\dot{\mathcal{E}}(E(t))$ denotes the time derivative of $\mathcal{E}(E)$ along the solution $E(t)$, $t \geq t_0$. Hence, it follows from (3.100) that

$$\mathcal{E}'(E)[w(E) - d(E) + S] \leq E^{\mathrm{T}}[S - d(E)], \quad E \in \overline{\mathbb{R}}_+^q, \quad S \in \mathbb{R}^q, \tag{3.101}$$

which implies that there exist continuous functions $\ell : \overline{\mathbb{R}}_+^q \to \mathbb{R}^p$ and $\mathcal{W} : \overline{\mathbb{R}}_+^q \to \mathbb{R}^{p \times q}$ such that

$$\begin{aligned} 0 = {} & \mathcal{E}'(E)[w(E) - d(E) + S] - E^{\mathrm{T}}[S - d(E)] \\ & + [\ell(E) + \mathcal{W}(E)S]^{\mathrm{T}}[\ell(E) + \mathcal{W}(E)S], \quad E \in \overline{\mathbb{R}}_+^q, \quad S \in \mathbb{R}^q. \end{aligned} \tag{3.102}$$

Now, equating coefficients of equal powers (of S), it follows that $\mathcal{W}(E) \equiv 0$, $\mathcal{E}'(E) = E^{\mathrm{T}}$, $E \in \overline{\mathbb{R}}_+^q$, and

$$0 = \mathcal{E}'(E)w(E) + \ell^{\mathrm{T}}(E)\ell(E), \quad E \in \overline{\mathbb{R}}_+^q. \tag{3.103}$$

Hence, $\mathcal{E}(E) = \frac{1}{2}E^{\mathrm{T}}E$, $E \in \overline{\mathbb{R}}_+^q$, and

$$0 = E^{\mathrm{T}}w(E) + \ell^{\mathrm{T}}(E)\ell(E), \quad E \in \overline{\mathbb{R}}_+^q. \tag{3.104}$$

Thus, (3.97) is a unique, continuously differentiable ectropy function for $\mathcal{G}$. $\square$

Note that it follows from the last equality in (3.99) that the ectropy function given by (3.97) satisfies (3.75) as an equality for an equilibrium process and as a strict inequality for a nonequilibrium process. Furthermore, it follows from (3.97) that ectropy is a measure of the extent to which the system energy deviates from a homogeneous state. Thus, ectropy is the dual of entropy and is a measure of the tendency of the large-scale dynamical system $\mathcal{G}$ to do useful work and grow more organized. Finally, we note that Theorem 3.5 can also be proven using Theorem 3.8 along with Theorem 2.7, where the inequality in (2.24) is reversed.

3.5 Semistability, Energy Equipartition, Irreversibility, and the Arrow of Time

Inequality (3.29) is a generalization of Clausius' inequality for equilibrium and nonequilibrium thermodynamics as well as reversible and irreversible thermodynamics as applied to large-scale dynamical systems, while inequality (3.75) is an anti–Clausius inequality. Moreover, for the ectropy function defined by (3.97), inequality (3.99) shows that a thermodynamically consistent large-scale dynamical system model is *dissipative* [100] with respect to the *supply rate* $E^{\mathrm{T}}S$ and with *storage function* corresponding to the system ectropy $\mathcal{E}(E)$. For the entropy function given by (3.61), note that $\mathcal{S}(0) = 0$ or, equivalently, $\lim_{E \to 0} \mathcal{S}(E) = 0$, which is consistent with the *third law of thermodynamics* (Nernst's theorem), which states that the entropy of every system at absolute zero can always be taken to be equal to zero.

For the (adiabatically) isolated large-scale dynamical system $\mathcal{G}$, (3.29) yields the fundamental inequality

$$\mathcal{S}(E(t_2)) \geq \mathcal{S}(E(t_1)), \quad t_2 \geq t_1. \tag{3.105}$$

Inequality (3.105) implies that, for any dynamical change in an adiabatically isolated large-scale dynamical system $\mathcal{G}$, the entropy of the final state can never be less than the entropy of the initial state. Inequality (3.105) is often identified with the second law of thermodynamics as a statement about entropy increase. It is important to stress that this result holds for an adiabatically isolated dynamical system. It is, however, possible with power (heat flux) supplied from an external system to reduce the entropy of the dynamical system $\mathcal{G}$. The entropy of both systems taken together, however, cannot decrease. These observations imply that when the isolated large-scale dynamical system $\mathcal{G}$ with thermodynamically consistent energy flow characteristics (i.e., Axioms i) and ii) hold) is at a state of maximum entropy consistent with its energy, it cannot be subject to any further dynamical change since any such change would result in a decrease of entropy. This of course implies that the state of *maximum entropy* is the stable state of an isolated system, and this equilibrium state has to be semistable.

Analogously, it follows from (3.75) that the isolated large-scale dynamical system $\mathcal{G}$ satisfies the fundamental inequality

$$\mathcal{E}(E(t_2)) \leq \mathcal{E}(E(t_1)), \quad t_2 \geq t_1, \tag{3.106}$$

which implies that the ectropy of the final state of $\mathcal{G}$ is always less than or equal to the ectropy of the initial state of $\mathcal{G}$. Hence, for the isolated large-scale dynamical system $\mathcal{G}$, the entropy increases if and only if the ectropy decreases. Thus, the state of *minimum ectropy* is the stable state of an isolated system, and this equilibrium state has to be semistable. This result can also be used to show that the isolated large-scale dynamical system $\mathcal{G}$ does not exhibit Poincaré recurrence. The next theorem concretizes the above observations.

Theorem 3.9 *Consider the large-scale dynamical system $\mathcal{G}$ with power balance equation (3.5) with $S(t) \equiv 0$ and $d(E) \equiv 0$, and assume that Axioms i) and ii) hold. Then for every $\alpha \geq 0$, αe is a semistable equilibrium state of (3.5). Furthermore, $E(t) \to \frac{1}{q}\mathbf{e}\mathbf{e}^{\mathrm{T}}E(t_0)$ as $t \to \infty$ and $\frac{1}{q}\mathbf{e}\mathbf{e}^{\mathrm{T}}E(t_0)$ is a semistable equilibrium state. Finally, if for some $k \in \{1, ..., q\}$, $\sigma_{kk}(E) \geq 0$, $E \in \overline{\mathbb{R}}_+^q$, and $\sigma_{kk}(E) = 0$ if and only if $E_k = 0$,[5] then the zero solution $E(t) \equiv 0$ to (3.5) is a globally asymptotically stable equilibrium state of (3.5).*

Proof. It follows from Axiom i) that $\alpha e \in \overline{\mathbb{R}}_+^q$, $\alpha \geq 0$, is an equilibrium state of (3.5). To show Lyapunov stability of the equilibrium state αe, consider the shifted-system ectropy function $\mathcal{E}_s(E) = \frac{1}{2}(E - \alpha e)^{\mathrm{T}}(E - \alpha e)$ as a Lyapunov function candidate. Now, since $\phi_{ij}(E) = -\phi_{ji}(E)$, $E \in \overline{\mathbb{R}}_+^q$, $i \neq j$, $i,j = 1, ..., q$, and $\mathbf{e}^{\mathrm{T}}w(E) = 0$, $E \in \overline{\mathbb{R}}_+^q$, it follows from Axiom ii) that

$$\dot{\mathcal{E}}_s(E) = (E - \alpha e)^{\mathrm{T}}\dot{E}$$
$$= (E - \alpha e)^{\mathrm{T}}w(E)$$

---

[5]The assumption $\sigma_{kk}(E) \geq 0$, $E \in \overline{\mathbb{R}}_+^q$, and $\sigma_{kk}(E) = 0$ if and only if $E_k = 0$ for some $k \in \{1, ..., q\}$ implies that if the kth subsystem possesses no energy, then this subsystem cannot dissipate energy to the environment. Conversely, if the kth subsystem does not dissipate energy to the environment, then this subsystem has no energy.

$$= E^{\mathrm{T}} w(E)$$

$$= \sum_{i=1}^{q} E_i \left[\sum_{j=1, j \neq i}^{q} \phi_{ij}(E) \right]$$

$$= \sum_{i=1}^{q} \sum_{j=i+1}^{q} (E_i - E_j) \phi_{ij}(E)$$

$$= \sum_{i=1}^{q} \sum_{j \in \mathcal{K}_i} (E_i - E_j) \phi_{ij}(E)$$

$$\leq 0, \quad E \in \overline{\mathbb{R}}_+^q, \qquad\qquad (3.107)$$

where $\mathcal{K}_i \triangleq \mathcal{N}_i \setminus \cup_{l=1}^{i-1}\{l\}$ and $\mathcal{N}_i \triangleq \{j \in \{1, ..., q\} : \phi_{ij}(E) = 0$ if and only if $E_i = E_j\}$, $i = 1, ..., q$, which establishes Lyapunov stability of the equilibrium state $\alpha \mathbf{e}$.

To show that $\alpha \mathbf{e}$ is semistable, let $\mathcal{R} \triangleq \{E \in \overline{\mathbb{R}}_+^q : \dot{\mathcal{E}}_{\mathrm{s}}(E) = 0\} = \{E \in \overline{\mathbb{R}}_+^q : (E_i - E_j)\phi_{ij}(E) = 0, i = 1, ..., q, j \in \mathcal{K}_i\}$. Now, by Axiom i) the directed graph associated with the connectivity matrix $\mathcal{C}$ for the large-scale dynamical system $\mathcal{G}$ is strongly connected, which implies that $\mathcal{R} = \{E \in \overline{\mathbb{R}}_+^q : E_1 = \cdots = E_q\}$. Since the set $\mathcal{R}$ consists of the equilibrium states of (3.5), it follows that the largest invariant set $\mathcal{M}$ contained in $\mathcal{R}$ is given by $\mathcal{M} = \mathcal{R}$. Hence, it follows from the Krasovskii-LaSalle invariant set theorem that for any initial condition $E(t_0) \in \overline{\mathbb{R}}_+^q$, $E(t) \to \mathcal{M}$ as $t \to \infty$, and hence, $\alpha \mathbf{e}$ is a semistable equilibrium state of (3.5). Next, note that since $\mathbf{e}^{\mathrm{T}} E(t) = \mathbf{e}^{\mathrm{T}} E(t_0)$ and $E(t) \to \mathcal{M}$ as $t \to \infty$, it follows that $E(t) \to \frac{1}{q} \mathbf{e}\mathbf{e}^{\mathrm{T}} E(t_0)$ as $t \to \infty$. Hence, with $\alpha = \frac{1}{q} \mathbf{e}^{\mathrm{T}} E(t_0)$, $\alpha \mathbf{e} = \frac{1}{q} \mathbf{e}\mathbf{e}^{\mathrm{T}} E(t_0)$ is a semistable equilibrium state of (3.5).

Finally, to show that in the case where for some $k \in \{1, ..., q\}$, $\sigma_{kk}(E) \geq 0$, $E \in \overline{\mathbb{R}}_+^q$, and $\sigma_{kk}(E) = 0$ if and only if $E_k = 0$, the zero solution $E(t) \equiv 0$ to (3.5) is globally asymptotically stable, consider the system ectropy $\mathcal{E}(E) = \frac{1}{2} E^{\mathrm{T}} E$, $E \in \overline{\mathbb{R}}_+^q$, as a candidate Lyapunov function. Note that $\mathcal{E}(0) = 0$, $\mathcal{E}(E) > 0$, $E \in \overline{\mathbb{R}}_+^q$, $E \neq 0$, and $\mathcal{E}(E)$ is radially unbounded. Now, the Lyapunov derivative along the system energy trajectories of (3.5) is given by

$$\dot{\mathcal{E}}(E) = E^{\mathrm{T}}[w(E) - d(E)]$$

$$= E^{\mathrm{T}} w(E) - E_k \sigma_{kk}(E)$$

$$= \sum_{i=1}^{q} E_i \left[\sum_{j=1, j \neq i}^{q} \phi_{ij}(E) \right] - E_k \sigma_{kk}(E)$$

$$= \sum_{i=1}^{q} \sum_{j=i+1}^{q} (E_i - E_j)\phi_{ij}(E) - E_k\sigma_{kk}(E)$$

$$= \sum_{i=1}^{q} \sum_{j \in \mathcal{K}_i} (E_i - E_j)\phi_{ij}(E) - E_k\sigma_{kk}(E)$$

$$\leq 0, \quad E \in \overline{\mathbb{R}}_+^q, \tag{3.108}$$

which shows that the zero solution $E(t) \equiv 0$ to (3.5) is Lyapunov stable. To show global asymptotic stability of the zero equilibrium state, let $\mathcal{R} \triangleq \{E \in \overline{\mathbb{R}}_+^q : \dot{\mathcal{E}}(E) = 0\} = \{E \in \overline{\mathbb{R}}_+^q : E_k\sigma_{kk}(E) = 0, \ k \in \{1,...,q\}\} \cap \{E \in \overline{\mathbb{R}}_+^q : (E_i - E_j)\phi_{ij}(E) = 0, \ i = 1,...,q, \ j \in \mathcal{K}_i\}$. Now, since Axiom i) holds and $\sigma_{kk}(E) = 0$ if and only if $E_k = 0$, it follows that $\mathcal{R} = \{E \in \overline{\mathbb{R}}_+^q : E_k = 0, \ k \in \{1,...,q\}\} \cap \{E \in \overline{\mathbb{R}}_+^q : E_1 = E_2 = \cdots = E_q\} = \{0\}$, and hence, the largest invariant set $\mathcal{M}$ contained in $\mathcal{R}$ is given by $\mathcal{M} = \{0\}$. Hence, it follows from the Krasovskii-LaSalle invariant set theorem that for any initial condition $E(t_0) \in \overline{\mathbb{R}}_+^q$, $E(t) \to \mathcal{M} = \{0\}$ as $t \to \infty$, which proves global asymptotic stability of the zero equilibrium state of (3.5). $\square$

Theorem 3.9 shows that the isolated (i.e., $S(t) \equiv 0$ and $d(E) \equiv 0$) large-scale dynamical system $\mathcal{G}$ is semistable. Hence, it follows from Theorem 2.16 that the isolated large-scale dynamical system $\mathcal{G}$ does not exhibit Poincaré recurrence in $\overline{\mathbb{R}}_+^q \setminus \mathcal{M}_e$. In Theorem 3.9 we used the shifted ectropy function to show that for the isolated (i.e., $S(t) \equiv 0$ and $d(E) \equiv 0$) large-scale dynamical system $\mathcal{G}$, $E(t) \to \frac{1}{q}\mathbf{e}\mathbf{e}^{\mathrm{T}}E(t_0)$ as $t \to \infty$ and $\frac{1}{q}\mathbf{e}\mathbf{e}^{\mathrm{T}}E(t_0)$ is a semistable equilibrium state. This result can also be arrived at using the system entropy. Specifically, using the system entropy given by (3.61), we can show attraction of the system trajectories to Lyapunov stable equilibrium points $\alpha\mathbf{e}$, $\alpha \geq 0$, and hence show semistability of these equilibrium states. To see this, note that since $\mathbf{e}^{\mathrm{T}}w(E) = 0$, $E \in \overline{\mathbb{R}}_+^q$, it follows that $\mathbf{e}^{\mathrm{T}}\dot{E}(t) = 0$, $t \geq t_0$. Hence, $\mathbf{e}^{\mathrm{T}}E(t) = \mathbf{e}^{\mathrm{T}}E(t_0)$, $t \geq t_0$. Furthermore, since $E(t) \geq\geq 0$, $t \geq t_0$, it follows that $0 \leq\leq E(t) \leq\leq \mathbf{e}\mathbf{e}^{\mathrm{T}}E(t_0)$, $t \geq t_0$, which implies that all solutions to (3.5) are bounded. Next, since by (3.63) the function $-\mathcal{S}(E(t))$, $t \geq t_0$, is nonincreasing and $E(t)$, $t \geq t_0$, is bounded, it follows from the Krasovskii-LaSalle invariant set theorem that for any initial condition $E(t_0) \in \overline{\mathbb{R}}_+^q$, $E(t) \to \mathcal{M}$ as $t \to \infty$, where $\mathcal{M}$ is the largest invariant set contained in $\mathcal{R} \triangleq \{E \in \overline{\mathbb{R}}_+^q : -\dot{\mathcal{S}}(E) = 0\}$. It now follows from the last inequality of (3.63) that $\mathcal{R} = \{E \in \overline{\mathbb{R}}_+^q : (E_i - E_j)\phi_{ij}(E) = 0, \ i = 1,...,q, \ j \in \mathcal{K}_i\}$, which, since the directed graph associated with the connectivity matrix $\mathcal{C}$

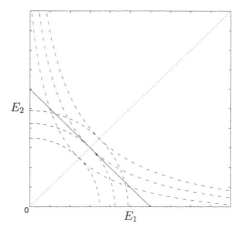

Figure 3.2 Thermodynamic equilibria $(\cdots)$, constant energy surfaces (——), constant ectropy surfaces $(--)$, and constant entropy surfaces $(-\cdot-)$.

for the large-scale dynamical system $\mathcal{G}$ is strongly connected, implies that $\mathcal{R} = \{E \in \overline{\mathbb{R}}_+^q : E_1 = \cdots = E_q\}$. Since the set $\mathcal{R}$ consists of the equilibrium states of (3.5), it follows that $\mathcal{M} = \mathcal{R}$, which, along with (3.107), establishes semistability of the equilibrium states $\alpha \mathbf{e}$, $\alpha \geq 0$.

Theorem 3.9 implies that the steady-state value of the energy in each subsystem $\mathcal{G}_i$ of the isolated large-scale dynamical system $\mathcal{G}$ is equal, that is, the steady-state energy of the isolated large-scale dynamical system $\mathcal{G}$ given by

$$E_\infty = \frac{1}{q} \mathbf{e} \mathbf{e}^{\mathrm{T}} E(t_0) = \left[\frac{1}{q} \sum_{i=1}^{q} E_i(t_0) \right] \mathbf{e} \qquad (3.109)$$

is uniformly distributed over all subsystems of $\mathcal{G}$. This phenomenon is known as *equipartition of energy*[6] [9, 10, 51, 69, 81] and is an emergent behavior in thermodynamic systems. The next proposition shows that, among all possible energy distributions in the large-scale dynamical system $\mathcal{G}$, energy equipartition corresponds to the minimum value of the system's ectropy and the maximum value of the system's entropy (see Figure 3.2).

Proposition 3.10 *Consider the large-scale dynamical system $\mathcal{G}$ with power balance equation (3.5), let $\mathcal{E} : \overline{\mathbb{R}}_+^q \to \overline{\mathbb{R}}_+$ and $\mathcal{S} : \overline{\mathbb{R}}_+^q \to \overline{\mathbb{R}}_+$ denote the ectropy and entropy functions of $\mathcal{G}$ given by (3.97) and*

---

[6]The phenomenon of equipartition of energy is closely related to the notion of a *monotemperaturic* system discussed in [17].

(3.61), respectively, and define $\mathcal{D}_c \triangleq \{E \in \overline{\mathbb{R}}_+^q : \mathbf{e}^T E = \beta\}$, where $\beta \geq 0$. Then

$$\arg\min_{E \in \mathcal{D}_c}(\mathcal{E}(E)) = \arg\max_{E \in \mathcal{D}_c}(\mathcal{S}(E)) = E^* = \frac{\beta}{q}\mathbf{e}. \qquad (3.110)$$

Furthermore, $\mathcal{E}_{\min} \triangleq \mathcal{E}(E^*) = \frac{1}{2}\frac{\beta^2}{q}$ and $\mathcal{S}_{\max} \triangleq \mathcal{S}(E^*) = q\log_e(c + \frac{\beta}{q}) - q\log_e c$.

Proof. The existence and uniqueness of E^* follows from the fact that $\mathcal{E}(E)$ and $-\mathcal{S}(E)$ are strictly convex continuous functions defined on the compact set $\mathcal{D}_c$. To minimize $\mathcal{E}(E) = \frac{1}{2}E^T E$ subject to $E \in \mathcal{D}_c$, form the Lagrangian $\mathcal{L}(E, \lambda) = \frac{1}{2}E^T E + \lambda(\mathbf{e}^T E - \beta)$, where $\lambda \in \mathbb{R}$ is a Lagrange multiplier. If E^* solves this minimization problem, then

$$0 = \left.\frac{\partial \mathcal{L}}{\partial E}\right|_{E=E^*} = E^{*T} + \lambda\mathbf{e}^T \qquad (3.111)$$

and hence $E^* = -\lambda\mathbf{e}$. Now, it follows from $\mathbf{e}^T E^* = \beta$ that $\lambda = -\frac{\beta}{q}$, which implies that $E^* = \frac{\beta}{q}\mathbf{e} \in \overline{\mathbb{R}}_+^q$. The fact that E^* minimizes the ectropy on the compact set $\mathcal{D}_c$ can be shown by computing the Hessian of the ectropy for the constrained parameter optimization problem and showing that the Hessian is positive definite at E^*. $\mathcal{E}_{\min} = \frac{1}{2}\frac{\beta^2}{q}$ is now immediate.

Analogously, to maximize $\mathcal{S}(E) = \mathbf{e}^T\log_e(c\mathbf{e} + E) - q\log_e c$ subject to $E \in \mathcal{D}_c$, form the Lagrangian $\mathcal{L}(E, \lambda) \triangleq \sum_{i=1}^q \log_e(c + E_i) + \lambda(\mathbf{e}^T E - \beta)$, where $\lambda \in \mathbb{R}$ is a Lagrange multiplier. If E^* solves this maximization problem, then

$$0 = \left.\frac{\partial \mathcal{L}}{\partial E}\right|_{E=E^*} = \left[\frac{1}{c+E_1^*} + \lambda, ..., \frac{1}{c+E_q^*} + \lambda\right]. \qquad (3.112)$$

Thus, $\lambda = -\frac{1}{c+E_i^*}$, $i = 1, ..., q$. If $\lambda = 0$, then the only value of E^* that satisfies (3.112) is $E^* = \infty$, which does not satisfy the constraint equation $\mathbf{e}^T E = \beta$ for finite $\beta \geq 0$. Hence, $\lambda \neq 0$ and $E_i^* = -(\frac{1}{\lambda} + c)$, $i = 1, ..., q$, which implies $E^* = -(\frac{1}{\lambda} + c)\mathbf{e}$. Now, it follows from $\mathbf{e}^T E^* = \beta$ that $-(\frac{1}{\lambda} + c) = \frac{\beta}{q}$ and hence $E^* = \frac{\beta}{q}\mathbf{e} \in \overline{\mathbb{R}}_+^q$. The fact that E^* maximizes the entropy on the compact set $\mathcal{D}_c$ can be shown by computing the Hessian of $\mathcal{S}(E)$ and showing that it is negative definite at E^*. $\mathcal{S}_{\max} = q\log_e(c + \frac{\beta}{q}) - q\log_e c$ is now immediate. $\square$

It follows from (3.105), (3.106), and Proposition 3.10 that conservation of energy in an isolated system necessarily implies nonconservation of ectropy and entropy. Hence, in an isolated large-scale dynamical system $\mathcal{G}$, all the energy, though always conserved, will eventually be degraded (diluted) to the point where it cannot produce any useful work. Hence, all motion would cease and the large-scale dynamical system would be fated to a state of eternal rest (semistability), wherein all subsystems will possess identical energies (energy equipartition). Ectropy would be a minimum and entropy would be a maximum giving rise to a state of absolute disorder. This is precisely what is known in theoretical physics as the *heat death of the universe*.[7]

Next, using the system entropy and ectropy functions given by (3.61) and (3.97), respectively, we show that our large-scale dynamical system $\mathcal{G}$ with power balance equation (3.5) is state irreversible for all nontrivial trajectories of $\mathcal{G}$ establishing a clear connection between our thermodynamic model and the arrow of time.

Theorem 3.10 *Consider the large-scale dynamical system $\mathcal{G}$ with power balance equation (3.5) with $S(t) \equiv 0$ and $d(E) \equiv 0$, and assume Axioms i) and ii) hold. Furthermore, let $s^E(\cdot, 0) \in \mathcal{W}_{[t_0, t_1]}$. Then for every $E_0 \notin \mathcal{M}_e$, there exists a continuously differentiable function $\mathcal{S} : \overline{\mathbb{R}}_+^q \to \mathbb{R}$ (respectively, $\mathcal{E} : \overline{\mathbb{R}}_+^q \to \mathbb{R}$) such that $\mathcal{S}(s^E(t, 0))$ (respectively, $\mathcal{E}(s^E(t, 0)))$ is a strictly increasing (respectively, decreasing) function of time. Furthermore, $s^E(\cdot, 0)$ is an I_q-reversible trajectory of $\mathcal{G}$ if and only if $s^E(t, 0) \in \mathcal{M}_e$, $t \in [t_0, t_1]$.*

Proof. The existence of a continuously differentiable function $\mathcal{S} : \overline{\mathbb{R}}_+^q \to \mathbb{R}$ (respectively, $\mathcal{E} : \overline{\mathbb{R}}_+^q \to \mathbb{R}$), which strictly increases (respectively, decreases) on all nontrivial trajectories of $\mathcal{G}$, is a restatement of Theorem 3.4 (respectively, Theorem 3.8) with $S(t) \equiv 0$ and $d(E) \equiv 0$. Now, necessity is immediate, while sufficiency is a direct consequence of Corollary 2.4 with $R = I_q$ and $V(z) = \mathcal{S}(E)$ (respectively, $V(z) = \mathcal{E}(E)$). $\square$

Theorem 3.10 shows that for every $E_0 \notin \mathcal{M}_e$, the adiabatically isolated dynamical system $\mathcal{G}$ is state irreversible. This gives a clear connection between our thermodynamic model and the arrow of time. In particular, it follows from Corollary 2.4 and Theorem 3.10 that there exists a function of the system state that strictly increases or strictly decreases in time on any nontrivial trajectory of $\mathcal{G}$ if and only if there does *not* exist a nontrivial reversible trajectory of $\mathcal{G}$. Thus, the existence of the continuously differentiable entropy and ectropy

---

[7]This *terroristic nimbus* of the second law of thermodynamics was first expressed in the work of Lord Kelvin in 1851 without any supporting mathematical arguments.

functions given by (3.61) and (3.97) for $\mathcal{G}$ establishes the existence of a completely ordered time set having a topological structure involving a closed set homeomorphic to the real line. This fact follows from the inverse function theorem of mathematical analysis and the fact that a continuous strictly monotonic function is a topological mapping (i.e., a homeomorphism), and conversely every topological mapping of a strictly monotonic function's domain onto its codomain must be strictly monotonic. This topological property gives a clear time-reversal asymmetry characterization of our thermodynamic model establishing an emergence of the direction of time flow.

We close this section by showing that our thermodynamically consistent large-scale system $\mathcal{G}$ satisfies *Gibbs' principle* [39, p. 56]. Gibbs' version of the second law of thermodynamics can be stated as follows:

> **Gibbs' Principle.** *For an equilibrium of any isolated system, it is necessary and sufficient that in all possible variations of the state of the system that do not alter its energy, the variation of its entropy shall either vanish or be negative.*

To establish Gibbs' principle for our thermodynamically consistent energy flow model, suppose $E_e = \alpha \mathbf{e}$, $\alpha \geq 0$, is an equilibrium state of the isolated system $\mathcal{G}$. Now, it follows from Proposition 3.10 that the entropy of $\mathcal{G}$ achieves its maximum at E_e subject to the constant energy level $\mathbf{e}^{\mathrm{T}} E = \alpha q$, $E \in \overline{\mathbb{R}}_+^q$. Hence, any variation of the state of the system that does not alter its energy leads to a zero or negative variation of the system entropy. Conversely, suppose that at some point $E^* \in \overline{\mathbb{R}}_+^q$ the variation of the system entropy is either zero or negative for all possible variations in the state of the system that do not alter the system's total energy. Furthermore, *ad absurdum*, let the isolated system $\mathcal{G}$ undergo an irreversible transformation starting at $E^* \notin \mathcal{M}_e$. Then, it follows from Theorem 3.4 that the entropy of $\mathcal{G}$ given by (3.61) strictly increases, which contradicts the above assumption. Hence, the system $\mathcal{G}$ cannot undergo an irreversible transformation starting at $E^* \notin \mathcal{M}_e$. Alternatively, if the isolated system $\mathcal{G}$ undergoes a reversible transformation starting at $E^* \in \mathcal{M}_e$, then E^* has to be an equilibrium state of $\mathcal{G}$.

Similarly, using the notion of ectropy, it can be shown that an isolated dynamical system $\mathcal{G}$ is in equilibrium if and only if, in all possible variations of the state of the system that do not alter its energy, the variation of the system ectropy is positive semidefinite. Finally, we note that a dual result to Gibbs' principle can be also established. Specifically, using similar arguments as outlined above, it can be shown that for an equilibrium point of any isolated system it is necessary and sufficient that, in all possible variations of the state of the system that do not alter its entropy (respectively, ectropy), the variation of its energy shall either vanish or be positive (respectively, negative).

3.6 Entropy Increase and the Second Law of Thermodynamics

In the preceding discussion it was assumed that our large-scale dynamical system model is such that energy flows from more energetic subsystems to less energetic subsystems, that is, heat (energy) flows in the direction of lower temperatures. Although this universal phenomenon can be predicted with virtual certainty, it follows as a manifestation of entropy and ectropy nonconservation for the case of two subsystems. To see this, consider the isolated large-scale dynamical system $\mathcal{G}$ with power balance equation (3.5) (with $S(t) \equiv 0$ and $d(E) \equiv 0$), and assume that the system entropy given by (3.61) is increasing and hence $\dot{\mathcal{S}}(E(t)) \geq 0$, $t \geq t_0$. Now, since

$$
\begin{aligned}
0 \leq \dot{\mathcal{S}}(E(t)) \\
&= \sum_{i=1}^{q} \frac{\dot{E}_i(t)}{c + E_i(t)} \\
&= \sum_{i=1}^{q} \sum_{j=1, j \neq i}^{q} \frac{\phi_{ij}(E(t))}{c + E_i(t)} \\
&= \sum_{i=1}^{q} \sum_{j=i+1}^{q} \left(\frac{\phi_{ij}(E(t))}{c + E_i(t)} - \frac{\phi_{ij}(E(t))}{c + E_j(t)} \right) \\
&= \sum_{i=1}^{q} \sum_{j \in \mathcal{K}_i} \frac{\phi_{ij}(E(t))(E_j(t) - E_i(t))}{(c + E_i(t))(c + E_j(t))}, \quad t \geq t_0, \qquad (3.113)
\end{aligned}
$$

it follows that for $q = 2$, $(E_1 - E_2)\phi_{12}(E) \leq 0$, $E \in \overline{\mathbb{R}}_+^2$, which implies that energy (heat) flows naturally from a more energetic subsystem (hot object) to a less energetic subsystem (cooler object). The universality of this emergent behavior thus follows from the fact that entropy (respectively, ectropy) transfer, accompanying energy transfer, always increases (respectively, decreases).

In the case where we have multiple subsystems, it is clear from (3.113) that entropy and ectropy nonconservation does not necessarily imply Axiom ii). However, if we invoke the additional condition (Axiom iii)) that if for any pair of connected subsystems $\mathcal{G}_k$ and $\mathcal{G}_l$, $k \neq l$, with energies $E_k \geq E_l$ (respectively, $E_k \leq E_l$) and for any other pair of connected subsystems $\mathcal{G}_m$ and $\mathcal{G}_n$, $m \neq n$, with energies $E_m \geq E_n$ (respectively, $E_m \leq E_n$), the inequality $\phi_{kl}(E)\phi_{mn}(E) \geq 0$, $E \in \overline{\mathbb{R}}_+^q$, holds, then nonconservation of entropy and ectropy in the isolated large-scale dynamical system $\mathcal{G}$ implies Axiom ii). The inequality $\phi_{kl}(E)\phi_{mn}(E) \geq 0$, $E \in \overline{\mathbb{R}}_+^q$, postulates that the direction of energy flow for any given pair of *energy similar* subsystems is consistent, that is, if for a given pair of connected subsystems at given different energy levels the energy flows in a certain direction, then for any other pair of connected subsystems with an analogous energy level difference, the energy flow direction is consistent with the original pair of subsystems. Note that this assumption does *not* specify the direction of energy flow between subsystems.

To see that $\dot{\mathcal{S}}(E(t)) \geq 0$, $t \geq t_0$, along with Axiom iii) implies Axiom ii), note that since (3.113) holds for all $t \geq t_0$ and $E(t_0) \in \overline{\mathbb{R}}_+^q$ is arbitrary, (3.113) implies

$$\sum_{i=1}^q \sum_{j \in \mathcal{K}_i} \frac{\phi_{ij}(E)(E_j - E_i)}{(c + E_i)(c + E_j)} \geq 0, \quad E \in \overline{\mathbb{R}}_+^q. \tag{3.114}$$

Now, it follows from (3.114) that for any fixed system energy level $E \in \overline{\mathbb{R}}_+^q$ there exists at least one pair of connected subsystems $\mathcal{G}_k$ and $\mathcal{G}_l$, $k \neq l$, such that $\phi_{kl}(E)(E_l - E_k) \geq 0$. Thus, if $E_k \geq E_l$ (respectively, $E_k \leq E_l$), then $\phi_{kl}(E) \leq 0$ (respectively, $\phi_{kl}(E) \geq 0$). Furthermore, it follows from Axiom iii) that for any other pair of connected subsystems $\mathcal{G}_m$ and $\mathcal{G}_n$, $m \neq n$, with $E_m \geq E_n$ (respectively, $E_m \leq E_n$) the inequality $\phi_{mn}(E) \leq 0$ (respectively, $\phi_{mn}(E) \geq 0$) holds, which implies that

$$\phi_{mn}(E)(E_n - E_m) \geq 0, \quad m \neq n. \tag{3.115}$$

Thus, it follows from (3.115) that energy (heat) flows naturally from more energetic subsystems (hot objects) to less energetic subsystems (cooler objects). Of course, since in the isolated large-scale dynamical system $\mathcal{G}$ ectropy decreases if and only if entropy increases, the same result can be arrived at by considering the ectropy of $\mathcal{G}$. Furthermore, since Axiom ii) holds, it follows from the conservation of energy and the fact that the large-scale dynamical system $\mathcal{G}$ is strongly connected that nonconservation of entropy and ectropy necessarily implies energy equipartition.

Finally, we close this section by showing that our definition of entropy given by (3.61) satisfies the eight criteria established in [46] for the acceptance of an analytic expression for representing a system entropy function. In particular, note that for a dynamical system $\mathcal{G}$: *i*) $\mathcal{S}(E)$ is well defined for every state $E \in \overline{\mathbb{R}}_+^q$ as long as $c > 0$. *ii*) If $\mathcal{G}$ is adiabatically isolated, then $\mathcal{S}(E(t))$ is a nondecreasing function of time. *iii*) If $\mathcal{S}_i(E_i) = \log_e(c + E_i) - \log_e c$ is the entropy of the *i*th subsystem of the system $\mathcal{G}$, then $\mathcal{S}(E) = \sum_{i=1}^q \mathcal{S}_i(E_i) = \mathbf{e}^T \log_e(c\mathbf{e} + E) - q \log_e c$, and hence, the system entropy $\mathcal{S}(E)$ is an additive quantity over all subsystems. *iv*) For the system $\mathcal{G}$, $\mathcal{S}(E) \geq 0$ for all $E \in \overline{\mathbb{R}}_+^q$. *v*) It follows from Proposition 3.10 that for a given value $\beta \geq 0$ of the total energy of the system $\mathcal{G}$, one and only one state, namely, $E^* = \frac{\beta}{q}\mathbf{e}$, corresponds to the largest value of $\mathcal{S}(E)$. *vi*) It follows from (3.61) that for the system $\mathcal{G}$, the graph of entropy versus energy is concave and smooth. *vii*) For a composite large-scale dynamical system $\mathcal{G}_C$ of two dynamical systems $\mathcal{G}_A$ and $\mathcal{G}_B$, the expression for the composite entropy $\mathcal{S}_C = \mathcal{S}_A + \mathcal{S}_B$, where $\mathcal{S}_A$ and $\mathcal{S}_B$ are entropies of $\mathcal{G}_A$ and $\mathcal{G}_B$, respectively, is such that the expression for the equilibrium state where the composite maximum entropy is achieved is identical to those obtained for $\mathcal{G}_A$ and $\mathcal{G}_B$ individually. Specifically, if q_A and q_B denote the number of subsystems in $\mathcal{G}_A$ and $\mathcal{G}_B$, respectively, and β_A and β_B denote the total energies of $\mathcal{G}_A$ and $\mathcal{G}_B$, respectively, then the maximum entropy of $\mathcal{G}_A$ and $\mathcal{G}_B$ individually is achieved at $E_A^* = \frac{\beta_A}{q_A}\mathbf{e}$ and $E_B^* = \frac{\beta_B}{q_B}\mathbf{e}$, respectively, while the maximum entropy of the composite system $\mathcal{G}_C$ is achieved at $E_C^* = \frac{\beta_A + \beta_B}{q_A + q_B}\mathbf{e}$. *viii*) It follows from Theorem 3.9 that for a stable equilibrium state $E = \frac{\beta}{q}\mathbf{e}$, where $\beta \geq 0$ is the total energy of the system $\mathcal{G}$ and q is the number of subsystems of $\mathcal{G}$, the entropy is totally defined by β and q, that is, $\mathcal{S}(E) = q \log_e(c + \frac{\beta}{q}) - q \log_e c$. Dual criteria to the eight criteria outlined above can also be established for an analytic expression representing system ectropy.

3.7 Interconnections of Thermodynamic Systems

In classical thermodynamics, it is not clear how the environment can be described in thermodynamic terms and is often not addressed by the theory. This is the case, for example, in the work of Carnot, Clausius, and Kelvin, wherein they consider the engine performing a cycle as the dynamical system and the reservoir belonging to the environment. This is in contrast to Planck, who views the reservoir as a thermodynamic system with a finite energy content and with the engine belonging to the environment. In this case, the thermodynamic state of the reservoir can change, and hence, the removal of energy from the reservoir via a cyclic engine need not be repeatable. Instead, the definition of a heat reservoir as formulated by Carnot and Kelvin is not at all transparent in the context of dynamical system theory since it can absorb or emit a finite amount of energy (heat) without changing its thermal (temperature) and deformation (volume) states, and hence, it possesses an infinite heat capacity. The issue is then whether the thermodynamic state of an infinite heat reservoir changes as it exchanges a finite quantity of heat with the dynamical system. This is not a trivial issue, and its assessment within a given theory has been one of the key demarcation points between physics and mathematics [96, p. 98]. Dynamical system theory makes no claim about the properties of any system larger than the system under consideration, and any exogenous effects are typically treated as external disturbances. Thus, the effect of the heat reservoir on the thermodynamic process can be viewed as an external disturbance to the system giving rise to an input-output open dynamical system.

In contrast, in classical mechanics it is always possible, at least in principle, to include interactions with the environment via *feedback* interconnecting components, consistent with Newton's third law, to obtain an augmented closed-loop feedback system. A feedback system consists of an interconnection of two systems, a *forward loop* system and a *feedback loop* system. The forward loop system is driven by an input and produces an output that serves as the input to the feedback loop system. The output of the feedback loop system, in turn, serves as the input to the forward loop system. Feedback systems are pervasive in nature and are ideal in capturing the behavior of interconnected dynamical systems.

To harmonize thermodynamics with mechanics, in this section we consider feedback interconnections of two thermodynamically consistent large-scale systems. This interconnection can correspond to a large-scale dynamical system with the environment or an interconnection between two large-scale dynamical systems. In the case where one of the systems corresponds to the environment, this formulation allows us to formally assign a state to the environment. In either case, we show that, under thermodynamically consistent assumptions on the feedback interconnection structure, the closed-loop system is guaranteed to be a thermodynamically consistent system, wherein the energy flow between the large-scale systems is due to the energy inflows and outflows of both systems.

Specifically, consider the large-scale dynamical system $\mathcal{G}$ with power balance equation

$$\dot{E}(t) = w(E(t)) - d(E(t)) + S(t), \quad E(t_0) = E_0, \quad t \geq t_0, \quad (3.116)$$

and outflow

$$y(t) = h(E(t)), \quad t \geq t_0, \quad (3.117)$$

and consider the large-scale dynamical system $\mathcal{G}_c$ with power balance equation

$$\dot{E}_c(t) = w_c(E_c(t)) - d_c(E_c(t)) + S_c(t), \quad E_c(t_0) = E_{c0}, \quad t \geq t_0, \quad (3.118)$$

and outflow

$$y_c(t) = h_c(E_c(t)), \quad t \geq t_0, \quad (3.119)$$

where $E(t) = [E_1(t), ..., E_q(t)]^T \in \overline{\mathbb{R}}_+^q, t \geq t_0, E_c(t) = [E_{c1}(t), ..., E_{cq_c}(t)]^T \in \overline{\mathbb{R}}_+^{q_c}, t \geq t_0, w \triangleq [w_1, ..., w_q]^T : \overline{\mathbb{R}}_+^q \rightarrow \mathbb{R}^q, w_i(E) = \sum_{j=1, j \neq i}^q [\sigma_{ij}(E) - \sigma_{ji}(E)], E \in \overline{\mathbb{R}}_+^q, i = 1, ..., q, w_c \triangleq [w_{c1}, ..., w_{cq_c}]^T : \overline{\mathbb{R}}_+^{q_c} \rightarrow \mathbb{R}^{q_c}, w_{ci}(E_c) = \sum_{j=1, j \neq i}^{q_c} [\sigma_{cij}(E_c) - \sigma_{cji}(E_c)], E_c \in \overline{\mathbb{R}}_+^{q_c}, i = 1, ..., q_c, d(E) \triangleq [\sigma_{11}(E), ..., \sigma_{qq}(E)]^T, E \in \overline{\mathbb{R}}_+^q, d_c(E_c) \triangleq [\sigma_{c11}(E_c), ..., \sigma_{cq_cq_c}(E_c)]^T, E_c \in \overline{\mathbb{R}}_+^{q_c}, S = [S_1, ..., S_q]^T : [0, \infty) \rightarrow \mathbb{R}^q, S_c = [S_{c1}, ..., S_{cq_c}]^T : [0, \infty) \rightarrow \mathbb{R}^{q_c}, h = [h_1, ..., h_{q_c}]^T : \overline{\mathbb{R}}_+^q \rightarrow \overline{\mathbb{R}}_+^{q_c}$, and $h_c = [h_{c1}, ..., h_{cq}]^T : \overline{\mathbb{R}}_+^{q_c} \rightarrow \overline{\mathbb{R}}_+^q$. To assure a feedback model consistent with energy conservation laws, the output power functions $h_i(\cdot), i = 1, ..., q_c$, and $h_{ci}(\cdot), i = 1, ..., q$, are given by

$$h_i(E) = \sum_{j=1}^q \eta_{ij}(E), \quad E \in \overline{\mathbb{R}}_+^q, \quad i = 1, ..., q_c, \quad (3.120)$$

and

$$h_{ci}(E_c) = \sum_{j=1}^{q_c} \eta_{cij}(E_c), \quad E_c \in \overline{\mathbb{R}}_+^{q_c}, \quad i = 1, ..., q, \quad (3.121)$$

where $\eta_{ij} : \overline{\mathbb{R}}_+^q \rightarrow \overline{\mathbb{R}}_+$ denotes the rate of energy flow from the jth subsystem of $\mathcal{G}$ to the ith subsystem of $\mathcal{G}_c$ and $\eta_{cij} : \overline{\mathbb{R}}_+^{q_c} \rightarrow \overline{\mathbb{R}}_+$ denotes the rate of energy flow from the jth subsystem of $\mathcal{G}_c$ to the ith subsystem of $\mathcal{G}$. The functions $\sigma_{ij} : \overline{\mathbb{R}}_+^q \rightarrow \overline{\mathbb{R}}_+, i, j = 1, ..., q$, have the same meaning and properties as defined in Section 3.2. The function $\sigma_{cij} : \overline{\mathbb{R}}_+^{q_c} \rightarrow \overline{\mathbb{R}}_+, i \neq j, i, j = 1, ..., q_c$, denotes the rate of energy flow from subsystem $\mathcal{G}_{cj}$ to subsystem $\mathcal{G}_{ci}$ of $\mathcal{G}_c$, and $\sigma_{cii} : \overline{\mathbb{R}}_+^{q_c} \to \overline{\mathbb{R}}_+$, $i = 1, ..., q_c$, denotes the rate of energy dissipation from the subsystem $\mathcal{G}_{ci}$.

In this case, conservation of energy implies that

$$\sigma_{ii}(E) = \sum_{j=1}^{q_c} \eta_{ji}(E), \quad E \in \overline{\mathbb{R}}_+^q, \quad i = 1, ..., q, \tag{3.122}$$

$$\sigma_{cii}(E_c) = \sum_{j=1}^{q} \eta_{cji}(E_c), \quad E_c \in \overline{\mathbb{R}}_+^{q_c}, \quad i = 1, ..., q_c, \tag{3.123}$$

and hence,

$$\sum_{i=1}^{q} \sigma_{ii}(E) = \mathbf{e}^T d(E) = \mathbf{e}_c^T h(E), \quad E \in \overline{\mathbb{R}}_+^q, \tag{3.124}$$

$$\sum_{i=1}^{q_c} \sigma_{cii}(E_c) = \mathbf{e}_c^T d_c(E_c) = \mathbf{e}^T h_c(E_c), \quad E_c \in \overline{\mathbb{R}}_+^{q_c}, \tag{3.125}$$

where $\mathbf{e}$ and $\mathbf{e}_c$ denote ones vectors of compatible dimensions. Here, we assume that $\sigma_{cij} : \overline{\mathbb{R}}_+^{q_c} \to \overline{\mathbb{R}}_+$, $i, j = 1, ..., q_c$, and $\eta_{cij} : \overline{\mathbb{R}}_+^{q_c} \to \overline{\mathbb{R}}_+$, $i = 1, ..., q$, $j = 1, ..., q_c$, are locally Lipschitz continuous on $\overline{\mathbb{R}}_+^{q_c}$, and $\sigma_{ij} : \overline{\mathbb{R}}_+^q \to \overline{\mathbb{R}}_+$, $i, j = 1, ..., q$, and $\eta_{ij} : \overline{\mathbb{R}}_+^q \to \overline{\mathbb{R}}_+$, $i = 1, ..., q_c$, $j = 1, ..., q$, are locally Lipschitz continuous on $\overline{\mathbb{R}}_+^q$.

Next, consider the positive feedback interconnection of the large-scale dynamical systems $\mathcal{G}$ and $\mathcal{G}_c$ shown in Figure 3.3 with $S = u + y_c$ and $S_c = u_c + y$, where $u \in \mathcal{U}$ and $u_c \in \mathcal{U}_{cc}$ are external power inflows to $\mathcal{G}$ and $\mathcal{G}_c$, respectively. Here, $\mathcal{U}$ is a subset of bounded continuous $\mathbb{R}^q$-valued functions on $\mathbb{R}$, and $\mathcal{U}_{cc}$ is a subset of bounded continuous $\mathbb{R}^{q_c}$-valued functions on $\mathbb{R}$. In this case, the closed-loop system dynamics are given by

$$\dot{\tilde{E}}(t) = \tilde{w}(\tilde{E}(t)) + \tilde{u}(t), \quad \tilde{E}(t_0) = \tilde{E}_0, \quad t \geq t_0, \tag{3.126}$$

where $\tilde{E}(t) \triangleq [E^T(t), E_c^T(t)]^T \in \overline{\mathbb{R}}_+^{\tilde{q}}$, $t \geq t_0$, $\tilde{q} \triangleq q + q_c$, $\tilde{u}(t) \triangleq [u^T(t), u_c^T(t)]^T \in \mathbb{R}^{\tilde{q}}$, $t \geq t_0$, and

$$\tilde{w}(\tilde{E}) \triangleq \begin{bmatrix} w(E) - d(E) + h_c(E_c) \\ w_c(E_c) - d_c(E_c) + h(E) \end{bmatrix}, \quad \tilde{E} \in \overline{\mathbb{R}}_+^{\tilde{q}}. \tag{3.127}$$

Note that the function $\tilde{w} : \overline{\mathbb{R}}_+^{\tilde{q}} \to \mathbb{R}^{\tilde{q}}$ is essentially nonnegative, and hence, it follows from Proposition 2.1 that the solution to (3.126) with

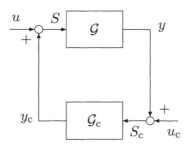

Figure 3.3 Feedback interconnection of large-scale systems $\mathcal{G}$ and $\mathcal{G}_c$.

$\tilde{u}(t) \equiv 0$ is nonnegative for all nonnegative initial conditions. Further-more, it follows from (3.124) and (3.125) that $\tilde{\mathbf{e}}^T \tilde{w}(\tilde{E}) = 0$, $\tilde{E} \in \overline{\mathbb{R}}_+^{\tilde{q}}$, where $\tilde{\mathbf{e}} \triangleq [\mathbf{e}^T, \mathbf{e}_c^T]^T$, and hence, energy is conserved in the isolated (i.e., $u(t) \equiv 0$ and $u_c(t) \equiv 0$) closed-loop feedback system consisting of $\mathcal{G}$ and $\mathcal{G}_c$. Next, let $\tilde{\phi}_{ij}(\tilde{E}) \triangleq \eta_{ij}(E) - \eta_{cji}(E_c)$, $\tilde{E} \in \overline{\mathbb{R}}_+^{\tilde{q}}$, denote the energy flow from the jth subsystem of $\mathcal{G}$ to the ith subsystem of $\mathcal{G}_c$. To assure that the closed-loop system possesses a thermodynamically consistent energy flow model, we require the following axioms on the interconnection structure for $\mathcal{G}$ and $\mathcal{G}_c$.

Axiom iv) $\mathcal{K} \triangleq \{(i,j) \in \{1, ..., q_c\} \times \{1, ..., q\} : \tilde{\phi}_{ij}(\tilde{E}) = 0$ *if and only if* $E_{ci} = E_j\} \neq \varnothing$, *and if* $(i,j) \notin \mathcal{K}$, *then* $\tilde{\phi}_{ij}(\tilde{E}) \equiv 0$.

Axiom v) *If* $(i,j) \in \mathcal{K}$, *then* $\tilde{\phi}_{ij}(\tilde{E})(E_{ci} - E_j) \leq 0$, $\tilde{E} \in \overline{\mathbb{R}}_+^{\tilde{q}}$.

Axiom iv) is analogous to Axiom i) and ensures that the dynamical systems $\mathcal{G}$ and $\mathcal{G}_c$ are strongly connected, that is, for any given ordered pair of subsystems $(\mathcal{G}_j, \mathcal{G}_{ci})$, $i = 1, ..., q_c$, $j = 1, ..., q$, there exists a path connecting this ordered pair. In other words, the rank of the connectivity matrix $\tilde{\mathcal{C}}$ associated with the feedback interconnection of $\mathcal{G}$ and $\mathcal{G}_c$ is equal to $\tilde{q} - 1$. Axiom v) is analogous to Axiom ii) and ensures that for any two connected subsystems $\mathcal{G}_j$ and $\mathcal{G}_{ci}$, $i = 1, ..., q_c$, $j = 1, ..., q$, such that $(i,j) \in \mathcal{K}$, energy flows from higher to lower energy levels. It follows from Axioms i) and iv) that $\tilde{E} = \alpha\tilde{\mathbf{e}} \in \overline{\mathbb{R}}_+^{\tilde{q}}$, $\alpha \geq 0$, is an equilibrium state of the isolated closed-loop system (3.126) given by the feedback interconnection of $\mathcal{G}$ and $\mathcal{G}_c$. The next theorem establishes that the equilibrium state $\alpha\tilde{\mathbf{e}} \in \overline{\mathbb{R}}_+^{\tilde{q}}$, $\alpha \geq 0$, of the isolated closed-loop system (3.126) is semistable.

Theorem 3.11 *Consider the positive feedback system consisting of* *the large-scale dynamical systems $\mathcal{G}$ and $\mathcal{G}_c$ given by (3.116)–(3.119) with $u(t) \equiv 0$ and $u_c(t) \equiv 0$, and assume that Axioms i), ii), iv), and v) hold. Then for every $\alpha \geq 0$, $\alpha \tilde{e} \in \overline{\mathbb{R}}_+^{\tilde{q}}$ is a semistable equilibrium state of (3.126). Furthermore, $\tilde{E}(t) \rightarrow \frac{1}{\tilde{q}} \tilde{e}(e^T E(t_0) + e_c^T E_c(t_0))$ as $t \rightarrow \infty$ and $\frac{1}{\tilde{q}} \tilde{e}(e^T E(t_0) + e_c^T E_c(t_0))$ is a semistable equilibrium state.*

Proof. It follows from Axioms i) and iv) that $\alpha \tilde{e} \in \overline{\mathbb{R}}_+^{\tilde{q}}$, $\alpha \geq 0$, is an equilibrium state of the isolated system (3.126). Next, consider the Lyapunov function candidate composed of the sum of the shifted ectropy functions of $\mathcal{G}$ and $\mathcal{G}_c$ given by

$$\tilde{\mathcal{E}}_s(\tilde{E}) = \tfrac{1}{2}(E - \alpha e)^T (E - \alpha e) + \tfrac{1}{2}(E_c - \alpha e_c)^T (E_c - \alpha e_c), \quad \tilde{E} \in \overline{\mathbb{R}}_+^{\tilde{q}}. \tag{3.128}$$

Computing the Lyapunov derivative of (3.128) and using Axioms i), ii), iv), and v), it follows that

$$\begin{aligned}
\dot{\tilde{\mathcal{E}}}_s(\tilde{E}) &= (E - \alpha e)^T [w(E) - d(E) + h_c(E_c)] \\
&\quad + (E_c - \alpha e_c)^T [w_c(E_c) - d_c(E_c) + h(E)] \\
&= \sum_{i=1}^{q} \sum_{j=i+1}^{q} \phi_{ij}(E)(E_i - E_j) + \sum_{i=1}^{q_c} \sum_{j=i+1}^{q_c} \phi_{cij}(E_c)(E_{ci} - E_{cj}) \\
&\quad + \sum_{i=1}^{q} \sum_{j=1}^{q_c} \tilde{\phi}_{ji}(\tilde{E})(E_{cj} - E_i) \\
&\leq 0, \quad \tilde{E} \in \overline{\mathbb{R}}_+^{\tilde{q}},
\end{aligned} \tag{3.129}$$

which establishes Lyapunov stability of the equilibrium state $\alpha \tilde{e} \in \overline{\mathbb{R}}_+^{\tilde{q}}$, $\alpha \geq 0$, of the feedback interconnection of $\mathcal{G}$ and $\mathcal{G}_c$.

Next, define the set $\mathcal{R} \triangleq \{\tilde{E} \in \overline{\mathbb{R}}_+^{\tilde{q}} : \dot{\tilde{\mathcal{E}}}_s(\tilde{E}) = 0\}$, and note that it follows from Axioms i), ii), iv), and v) that $\mathcal{R} = \{\tilde{E} \in \overline{\mathbb{R}}_+^{\tilde{q}} : E_1 = \cdots = E_q\} \cap \{\tilde{E} \in \overline{\mathbb{R}}_+^{\tilde{q}} : E_{c1} = \cdots = E_{cq_c}\} \cap \{\tilde{E} \in \overline{\mathbb{R}}_+^{\tilde{q}} : E_{cj} = E_i, (j, i) \in \mathcal{K}\} = \{\tilde{E} \in \overline{\mathbb{R}}_+^{\tilde{q}} : E_1 = \cdots = E_q = E_{c1} = \cdots = E_{cq_c}\}$. Since $\mathcal{R}$ consists of only equilibrium states of the feedback interconnection of $\mathcal{G}$ and $\mathcal{G}_c$, it follows that the largest invariant set $\mathcal{M}$ contained in $\mathcal{R}$ is given by $\mathcal{M} = \mathcal{R}$. Thus, it follows from the Krasovskii-LaSalle invariant set theorem that for any initial condition $\tilde{E}(t_0) \in \overline{\mathbb{R}}_+^{\tilde{q}}$, $\tilde{E}(t) \rightarrow \mathcal{M}$ as $t \rightarrow \infty$, and hence, $\alpha \tilde{e} \in \overline{\mathbb{R}}_+^{\tilde{q}}$ is a semistable equilibrium state of the isolated system (3.126). Finally, note that since $(e^T E(t) + e_c^T E_c(t)) = (e^T E(t_0) + e_c^T E_c(t_0))$, $t \geq t_0$, and $\tilde{E}(t) \rightarrow \mathcal{M}$ as $t \rightarrow \infty$, it follows that $\tilde{E}(t) \to \frac{1}{\tilde{q}}\tilde{\mathbf{e}}(\mathbf{e}^{\mathrm{T}}E(t_0) + \mathbf{e}_{\mathrm{c}}^{\mathrm{T}}E_{\mathrm{c}}(t_0))$ as $t \to \infty$. Hence, with $\alpha = \frac{1}{\tilde{q}}(\mathbf{e}^{\mathrm{T}}E(t_0) + \mathbf{e}_{\mathrm{c}}^{\mathrm{T}}E_{\mathrm{c}}(t_0))$, $\alpha\tilde{\mathbf{e}} = \frac{1}{\tilde{q}}\tilde{\mathbf{e}}(\mathbf{e}^{\mathrm{T}}E(t_0) + \mathbf{e}_{\mathrm{c}}^{\mathrm{T}}E_{\mathrm{c}}(t_0))$ is a semistable equilibrium state of the isolated closed-loop system (3.126).   □

The entropy and ectropy functions for the closed-loop system (3.126) can be constructed by appropriately combining the entropy and ectropy functions of $\mathcal{G}$ and $\mathcal{G}_{\mathrm{c}}$. In particular, the ectropy of the closed-loop system (3.126) is the sum of the individual ectropy functions of $\mathcal{G}$ and $\mathcal{G}_{\mathrm{c}}$ and is given by

$$\tilde{\mathcal{E}}(\tilde{E}) = \mathcal{E}(E) + \mathcal{E}_{\mathrm{c}}(E_{\mathrm{c}}) = \tfrac{1}{2}E^{\mathrm{T}}E + \tfrac{1}{2}E_{\mathrm{c}}^{\mathrm{T}}E_{\mathrm{c}} = \tfrac{1}{2}\tilde{E}^{\mathrm{T}}\tilde{E}, \quad \tilde{E} \in \overline{\mathbb{R}}_{+}^{\tilde{q}}. \tag{3.130}$$

Hence, it follows that

$$\dot{\tilde{\mathcal{E}}}(\tilde{E}) = \sum_{i=1}^{q}\sum_{j=i+1}^{q} \phi_{ij}(E)(E_i - E_j) + \sum_{i=1}^{q_{\mathrm{c}}}\sum_{j=i+1}^{q_{\mathrm{c}}} \phi_{\mathrm{c}ij}(E_{\mathrm{c}})(E_{\mathrm{c}i} - E_{\mathrm{c}j})$$
$$+ \sum_{i=1}^{q}\sum_{j=1}^{q_{\mathrm{c}}} \tilde{\phi}_{ji}(\tilde{E})(E_{\mathrm{c}j} - E_i)$$
$$\leq 0, \quad \tilde{E} \in \overline{\mathbb{R}}_{+}^{\tilde{q}}, \tag{3.131}$$

which shows that $\tilde{\mathcal{E}}(\tilde{E})$ is a nonincreasing function of time. Note that since the closed-loop system (3.126) satisfies Axioms i), ii), iv), and v), it follows from Theorem 3.8 that $\tilde{\mathcal{E}}(\tilde{E})$, $\tilde{E} \in \overline{\mathbb{R}}_{+}^{\tilde{q}}$, is a unique, continuously differentiable ectropy function of (3.126).

Similarly, the entropy of the isolated closed-loop system (3.126) is the sum of the individual entropy functions of $\mathcal{G}$ and $\mathcal{G}_{\mathrm{c}}$ and is given by

$$\begin{aligned}\tilde{\mathcal{S}}(\tilde{E}) &= \mathcal{S}(E) + \mathcal{S}_{\mathrm{c}}(E_{\mathrm{c}})\\ &= \mathbf{e}^{\mathrm{T}}\mathbf{log}(c\mathbf{e} + E) + \mathbf{e}_{\mathrm{c}}^{\mathrm{T}}\mathbf{log}(c\mathbf{e}_{\mathrm{c}} + E_{\mathrm{c}}) - \tilde{q}\log_e c, \quad \tilde{E} \in \overline{\mathbb{R}}_{+}^{\tilde{q}}.\end{aligned} \tag{3.132}$$

Hence, it follows that

$$\dot{\tilde{\mathcal{S}}}(\tilde{E}) = \sum_{i=1}^{q}\sum_{j=i+1}^{q} \frac{\phi_{ij}(E)(E_j - E_i)}{(c + E_i)(c + E_j)} + \sum_{i=1}^{q_{\mathrm{c}}}\sum_{j=i+1}^{q_{\mathrm{c}}} \frac{\phi_{\mathrm{c}ij}(E_{\mathrm{c}})(E_{\mathrm{c}j} - E_{\mathrm{c}i})}{(c + E_{\mathrm{c}i})(c + E_{\mathrm{c}j})}$$
$$+ \sum_{i=1}^{q}\sum_{j=1}^{q_{\mathrm{c}}} \frac{\tilde{\phi}_{ji}(\tilde{E})(E_i - E_{\mathrm{c}j})}{(c + E_i)(c + E_{\mathrm{c}j})}$$
$$\geq 0, \quad \tilde{E} \in \overline{\mathbb{R}}_{+}^{\tilde{q}}. \tag{3.133}$$

As in the case of ectropy, it follows from Theorem 3.4 that $\tilde{S}(\tilde{E})$, $\tilde{E} \in$ $\overline{\mathbb{R}}_+^{\tilde{q}}$, is a unique, continuously differentiable entropy function for the closed-loop system (3.126). Finally, note that it follows from Axioms i), ii), iv), and v) that the entropy (respectively, ectropy) of the closed-loop system (3.126) given by (3.132) (respectively, (3.130)) satisfies (3.133) (respectively, (3.131)) as a strict inequality for $\tilde{E}(t) \notin$ $\mathcal{M}_e$, $t \geq t_0$, and as an equality for $\tilde{E}(t) \in \mathcal{M}_e$, $t \geq t_0$, where $\mathcal{M}_e = \{\tilde{E} \in \overline{\mathbb{R}}_+^{\tilde{q}} : \alpha \tilde{\mathbf{e}}, \alpha \geq 0\}$. Hence, it follows from Theorem 3.10, as applied to the isolated closed-loop system (3.126), that (3.126) is state irreversible for all nontrivial trajectories.

3.8 Monotonicity of System Energies in Thermodynamic Processes

Even though Theorem 3.9 gives sufficient conditions under which the subsystem energies in the large-scale dynamical system $\mathcal{G}$ converge, these subsystem energies may exhibit an oscillatory (hyperbolic) or nonmonotonic behavior prior to convergence. For certain thermodynamical processes, it is desirable to identify system models that guarantee monotonicity of the system energy flows. It is important to note that monotonicity of solutions does not necessarily imply Axiom ii), nor does Axiom ii) imply monotonicity of solutions. These are two disjoint notions. In this section, we give necessary and sufficient conditions under which the solutions to (3.5) are monotonic.

To develop necessary and sufficient conditions for monotonicity of solutions, note that the power balance equation (3.5) for the large-scale dynamical system $\mathcal{G}$ can be written as

$$\dot{E}(t) = [\mathcal{J}(E(t)) - \mathcal{D}(E(t))] \left(\frac{\partial \mathcal{H}}{\partial E}(E(t)) \right)^{\mathrm{T}} + GS(t), \quad E(t_0) = E_0,$$

$$t \geq t_0, \quad (3.134)$$

where $E(t) \in \overline{\mathbb{R}}_+^q$, $\mathcal{H}(E) = \mathbf{e}^{\mathrm{T}} E$, $S(t) = [S_1(t), ..., S_q(t)]^{\mathrm{T}}$, $t \geq t_0$, $\mathcal{J}(E)$ is a skew-symmetric matrix function with $\mathcal{J}_{(i,i)}(E) = 0$ and $\mathcal{J}_{(i,j)}(E) = \sigma_{ij}(E) - \sigma_{ji}(E)$, $i \neq j$, $i, j = 1, ..., q$, $\mathcal{D}(E) = \mathrm{diag}[\sigma_{11}(E), ..., \sigma_{qq}(E)] \geq 0$, and $G \in \mathbb{R}^{q \times q}$ is a diagonal input matrix that has been included for generality and contains zeros and ones as its entries. Hence, the power balance equation of the large-scale dynamical system $\mathcal{G}$ has a *port-controlled Hamiltonian* structure [67] with a Hamiltonian function $\mathcal{H}(E) = \mathbf{e}^{\mathrm{T}} E = \sum_{i=1}^q E_i$ representing the sum of all subsystem energies, $\mathcal{D}(E)$ representing power dissipation in the subsystems, $\mathcal{J}(E) = -\mathcal{J}^{\mathrm{T}}(E)$ representing energy-conserving subsystem coupling, and $S(t)$, $t \geq t_0$, representing supplied system power. As noted in Section 3.3, the nonlinear power balance equation (3.134) can exhibit a full range of nonlinear behavior, including bifurcations, limit cycles, and even chaos. However, a thermodynamically consistent energy flow model ensures that the evolution of the system energy is diffusive in character with convergent subsystem energies. As shown in Section 3.3, Axioms i) and ii) guarantee a thermodynamically consistent energy flow model.

In order to guarantee a thermodynamically consistent energy flow model, we assume Axiom ii) holds and seek solutions to (3.134) that exhibit a monotonic behavior of the subsystem energies. This would physically imply that the energy of a subsystem whose initial energy is greater than the average system energy will decrease, while the energy of a subsystem whose initial energy is less than the average system energy will increase. This of course is consistent with the second law of thermodynamics with the additional constraint of monotonic heat flows. The following definition is needed.

Definition 3.4 *Consider the large-scale dynamical system $\mathcal{G}$ with power balance equation (3.134). The subsystem energies $E(t)$, $t \geq t_0$, of $\mathcal{G}$ are monotonic for all $E_0 \in \mathcal{D}_c \subseteq \overline{\mathbb{R}}_+^q$, where $\mathcal{D}_c$ is a positively invariant set with respect to (3.134), if there exists a weighting matrix $R \in \mathbb{R}^{q \times q}$ such that $R = \mathrm{diag}[r_1, ..., r_q]$, $r_i = \pm 1$, $i = 1, ..., q$, and for every $E_0 \in \mathcal{D}_c \subseteq \overline{\mathbb{R}}_+^q$, $RE(t_2) \leq\leq RE(t_1)$, $t_0 \leq t_1 \leq t_2$.*

The following result presents necessary and sufficient conditions that guarantee that the subsystem energies of the large-scale dynamical system $\mathcal{G}$ are monotonic. It is important to note that this result holds whether or not Axiom ii) holds.

Theorem 3.12 *Consider the large-scale dynamical system $\mathcal{G}$ with power balance equation (3.134). Then the following statements hold:*

    i) *If $S(t) \geq\geq 0$, $t \geq t_0$, and there exists a matrix $R \in \mathbb{R}^{q \times q}$ such that $R = \mathrm{diag}[r_1, ..., r_q]$, $r_i = \pm 1$, $i = 1, ..., q$, $R[\mathcal{J}(E) - \mathcal{D}(E)](\frac{\partial \mathcal{H}}{\partial E}(E))^{\mathrm{T}} \leq\leq 0$, $E \in \overline{\mathbb{R}}_+^q$, and $RG \leq\leq 0$, then the subsystem energies $E(t)$, $t \geq t_0$, of $\mathcal{G}$ are monotonic for all $E_0 \in \overline{\mathbb{R}}_+^q$.*

    ii) *Let $S(t) \equiv 0$ and let $\mathcal{D}_c \subseteq \overline{\mathbb{R}}_+^q$ be a positively invariant set with respect to (3.134). Then the subsystem energies $E(t)$, $t \geq t_0$, of $\mathcal{G}$ are monotonic for all $E_0 \in \mathcal{D}_c \subseteq \overline{\mathbb{R}}_+^q$ if and only if there exists a matrix $R \in \mathbb{R}^{q \times q}$ such that $R = \mathrm{diag}[r_1, ..., r_q]$, $r_i = \pm 1$, $i = 1, ..., q$, and $R[\mathcal{J}(E) - \mathcal{D}(E)](\frac{\partial \mathcal{H}}{\partial E}(E))^{\mathrm{T}} \leq\leq 0$, $E \in \mathcal{D}_c \subseteq \overline{\mathbb{R}}_+^q$.*

Proof. $i)$ Let $S(t) \geq\geq 0$, $t \geq t_0$, and assume there exists $R = \text{diag}[r_1, ..., r_q]$, $r_i = \pm 1$, $i = 1, ..., q$, such that $R[\mathcal{J}(E) - \mathcal{D}(E)]$ $\cdot (\frac{\partial \mathcal{H}}{\partial E}(E))^{\mathrm{T}} \leq\leq 0$, $E \in \overline{\mathbb{R}}_+^q$. Now, it follows from (3.134) that

$$R\dot{E}(t) = R[\mathcal{J}(E(t)) - \mathcal{D}(E(t))] \left(\frac{\partial \mathcal{H}}{\partial E}(E(t)) \right)^{\mathrm{T}} + RGS(t),$$
$$E(t_0) = E_0, \quad t \geq t_0, \quad (3.135)$$

which further implies that

$$RE(t_2) = RE(t_1) + \int_{t_1}^{t_2} R[\mathcal{J}(E(t)) - \mathcal{D}(E(t))] \left(\frac{\partial \mathcal{H}}{\partial E}(E(t)) \right)^{\mathrm{T}} \mathrm{d}t$$
$$+ \int_{t_1}^{t_2} RGS(t)\mathrm{d}t. \quad (3.136)$$

Next, since $[\mathcal{J}(E) - \mathcal{D}(E)](\frac{\partial \mathcal{H}}{\partial E}(E))^{\mathrm{T}}$ is essentially nonnegative and $S(t) \geq\geq 0$, $t \geq t_0$, it follows from Proposition 3.1 that $E(t) \geq\geq 0$, $t \geq t_0$, for all $E_0 \in \overline{\mathbb{R}}_+^q$. Hence, since $R[\mathcal{J}(E) - \mathcal{D}(E)](\frac{\partial \mathcal{H}}{\partial E}(E))^{\mathrm{T}} \leq\leq 0$, $E \in \overline{\mathbb{R}}_+^q$, and $RG \leq\leq 0$, it follows that

$$R[\mathcal{J}(E(t)) - \mathcal{D}(E(t))](\frac{\partial \mathcal{H}}{\partial E}(E(t)))^{\mathrm{T}} + RGS(t) \leq\leq 0, \quad t \geq t_0,$$
$$(3.137)$$

which implies that for every $E_0 \in \overline{\mathbb{R}}_+^q$, $RE(t_2) \leq\leq RE(t_1)$, $t_0 \leq t_1 \leq t_2$.

$ii)$ To show sufficiency, note that since by assumption $\mathcal{D}_c$ is positively invariant, then $R[\mathcal{J}(E(t)) - \mathcal{D}(E(t))](\frac{\partial \mathcal{H}}{\partial E}(E(t)))^{\mathrm{T}} \leq\leq 0$, $t \geq t_0$, for all $E_0 \in \mathcal{D}_c \subseteq \overline{\mathbb{R}}_+^q$. Now, the result follows by using identical arguments as in $i)$ with $S(t) \equiv 0$ and $E_0 \in \mathcal{D}_c \subseteq \overline{\mathbb{R}}_+^q$. To show necessity, assume that (3.134) with $S(t) \equiv 0$ is monotonic for all $E_0 \in \mathcal{D}_c \subseteq \overline{\mathbb{R}}_+^q$. In this case, (3.135) implies that for every $\tau > t_0$,

$$RE(\tau) = RE_0 + \int_{t_0}^{\tau} R[\mathcal{J}(E(t)) - \mathcal{D}(E(t))] \left(\frac{\partial \mathcal{H}}{\partial E}(E(t)) \right)^{\mathrm{T}} \mathrm{d}t. \quad (3.138)$$

Now, suppose, *ad absurdum*, there exist $J \in \{1, ..., q\}$ and $E_0 \in \mathcal{D}_c \subseteq \overline{\mathbb{R}}_+^q$ such that $[R[\mathcal{J}(E_0) - \mathcal{D}(E_0)](\frac{\partial \mathcal{H}}{\partial E}(E_0))^{\mathrm{T}}]_J > 0$. Since the mapping $R[\mathcal{J}(\cdot) - \mathcal{D}(\cdot)](\frac{\partial \mathcal{H}}{\partial E}(\cdot))^{\mathrm{T}}$ and the solution $E(t)$, $t \geq t_0$, to (3.134) are continuous, it follows that there exists $\tau > t_0$ such that

$$[R[\mathcal{J}(E(t)) - \mathcal{D}(E(t))](\frac{\partial \mathcal{H}}{\partial E}(E(t)))^{\mathrm{T}}]_J > 0, \quad t_0 \leq t \leq \tau, \quad (3.139)$$

which implies that $[RE(\tau)]_J > [RE_0]_J$, leading to a contradiction. Hence, $R[\mathcal{J}(E) - \mathcal{D}(E)](\frac{\partial \mathcal{H}}{\partial E}(E))^{\mathrm{T}} \leq\leq 0$, $E \in \mathcal{D}_c \subseteq \overline{\mathbb{R}}_+^q$. $\square$

It follows from i) of Theorem 3.12 that if $G = I_q$ (that is, external power (heat flux) can be injected to all subsystems), then $R = -I_q$, and hence, $[\mathcal{J}(E) - \mathcal{D}(E)](\frac{\partial \mathcal{H}}{\partial E}(E))^{\mathrm{T}} \geq\geq 0$, $E \in \overline{\mathbb{R}}_+^q$. This case would correspond to a power balance equation whose states are all increasing and can only be achieved if $\mathcal{D}(E) = 0$, $E \in \overline{\mathbb{R}}_+^q$. This, of course, implies that the dynamical system $\mathcal{G}$ cannot dissipate energy, and hence, the transfer of energy (heat) from a lower energy (temperature) level (source) to a higher energy (temperature) level (sink) requires the input of additional heat or energy. This is consistent with Clausius' statement of the second law of thermodynamics.

The following result is a direct consequence of Theorem 3.12 and provides sufficient conditions for convergence of the subsystem energies of the isolated large-scale dynamical system $\mathcal{G}$. Once again, this result holds whether or not Axiom ii) holds.

Theorem 3.13 *Consider the large-scale dynamical system $\mathcal{G}$ with power balance equation (3.134) and $S(t) \equiv 0$. Let $\mathcal{D}_c \subseteq \overline{\mathbb{R}}_+^q$ be a positively invariant set. If there exists a matrix $R \in \mathbb{R}^{q \times q}$ such that $R =$ $\mathrm{diag}[r_1, ..., r_q]$, $r_i = \pm 1$, $i = 1, ..., q$, and $R[\mathcal{J}(E) - \mathcal{D}(E)](\frac{\partial \mathcal{H}}{\partial E}(E))^{\mathrm{T}}$ $\leq\leq 0$, $E \in \mathcal{D}_c \subseteq \overline{\mathbb{R}}_+^q$, then for every $E_0 \in \mathcal{D}_c \subseteq \overline{\mathbb{R}}_+^q$, $\lim_{t \to \infty} E(t)$ exists.*

Proof. Since $\mathcal{H}(E) = \mathbf{e}^{\mathrm{T}} E$, $E \in \overline{\mathbb{R}}_+^q$, it follows that

$$
\begin{aligned}
\dot{\mathcal{H}}(E) &= \frac{\partial \mathcal{H}}{\partial E} \dot{E} \\
&= \frac{\partial \mathcal{H}}{\partial E}[\mathcal{J}(E) - \mathcal{D}(E)](\frac{\partial \mathcal{H}}{\partial E})^{\mathrm{T}} \\
&= -\frac{\partial \mathcal{H}}{\partial E} \mathcal{D}(E)(\frac{\partial \mathcal{H}}{\partial E})^{\mathrm{T}} \leq 0, \quad E \in \overline{\mathbb{R}}_+^q,
\end{aligned} \tag{3.140}
$$

and hence, $\dot{\mathcal{H}}(E(t)) \leq 0$, $t \geq t_0$, where $E(t)$, $t \geq t_0$, denotes the solution of (3.134). This implies that $\mathcal{H}(E(t)) \leq \mathcal{H}(E_0) = \mathbf{e}^{\mathrm{T}} E_0$, $t \geq t_0$, and hence, for every $E_0 \in \overline{\mathbb{R}}_+^q$, the solution $E(t)$, $t \geq t_0$, of (3.134) is bounded. Hence, for every $i \in \{1, ..., q\}$, $E_i(t)$, $t \geq t_0$, is bounded. Furthermore, it follows from Theorem 3.12 that $E_i(t)$, $t \geq t_0$, is monotonic for all $E_0 \in \mathcal{D}_c \subseteq \overline{\mathbb{R}}_+^q$. Now, since $E_i(\cdot)$, $i \in \{1, ..., q\}$, is continuous and every bounded nonincreasing or nondecreasing scalar sequence converges to a finite real number, it follows from the monotone convergence theorem that $\lim_{t \to \infty} E_i(t)$, $i \in \{1, ..., q\}$, exists. Hence, $\lim_{t \to \infty} E(t)$ exists for all $E_0 \in \mathcal{D}_c \subseteq \overline{\mathbb{R}}_+^q$. $\square$

Chapter Four

Temperature Equipartition and the Kinetic Theory of Gases

4.1 Semistability and Temperature Equipartition

The thermodynamic axioms introduced in Chapter 3 postulate that subsystem energies are synonymous with subsystem temperatures. In this chapter, we generalize the results of Chapter 3 to the case where the subsystem energies are proportional to the subsystem temperatures with the proportionality constants representing the subsystem *specific heats* or *thermal capacities*.[1] In the case where the specific heats of all the subsystems are equal, the results of this section specialize to those of Chapter 3. To include temperature notions in our large-scale dynamical system model, we replace Axioms i) and ii) of Section 3.3 with the following axioms. Let $\beta_i > 0$, $i = 1, \ldots, q$, denote the reciprocal of the specific heat (at constant volume) of the ith subsystem $\mathcal{G}_i$ so that the *absolute temperature* in the ith subsystem is given by $\hat{T}_i = \beta_i E_i$.

Axiom i) *For the connectivity matrix $\mathcal{C} \in \mathbb{R}^{q \times q}$ associated with the large-scale dynamical system $\mathcal{G}$ defined by (3.24) and (3.25), rank $\mathcal{C}$ $= q{-}1$, and for $\mathcal{C}_{(i,j)} = 1$, $i \neq j$, $\phi_{ij}(E) = 0$ if and only if $\beta_i E_i = \beta_j E_j$.*

Axiom ii) *For $i, j = 1, \ldots, q$, $(\beta_i E_i - \beta_j E_j)\phi_{ij}(E) \leq 0$, $E \in \overline{\mathbb{R}}_+^q$.*

Axiom i) implies that if the temperatures in the connected subsystems $\mathcal{G}_i$ and $\mathcal{G}_j$ are equal, then heat exchange between these subsystems is not possible. This is a statement of the *zeroth law of thermodynamics*, which postulates that temperature equality is a necessary and sufficient condition for *thermal equilibrium*. Axiom ii) implies that heat (energy) must flow in the direction of lower temperatures. This is a statement of the *second law of thermodynamics*, which states

---

[1]The *thermal capacity* of a body is the ratio of the infinitesimal amount of heat absorbed by the body to the infinitesimal increase in temperature produced by this heat. In general, the thermal capacity of a body is different if the body is heated at a constant volume or at a constant pressure.

that a transformation whose only final result is to transfer heat from a body at a given temperature to a body at a higher temperature is impossible. The following proposition is needed for the statement of the main results of this section.

Proposition 4.1 *Consider the large-scale dynamical system $\mathcal{G}$ with power balance equation (3.5), and assume that Axioms i) and ii) hold. Then for all $E_0 \in \overline{\mathbb{R}}_+^q$, $t_f \geq t_0$, and $S(\cdot) \in \mathcal{U}$ such that $E(t_f) = E(t_0) = E_0$,*

$$\int_{t_0}^{t_f} \sum_{i=1}^{q} \frac{S_i(t) - \sigma_{ii}(E(t))}{c + \beta_i E_i(t)} dt = \oint \sum_{i=1}^{q} \frac{dQ_i(t)}{c + \beta_i E_i(t)} \leq 0 \qquad (4.1)$$

and

$$\int_{t_0}^{t_f} \sum_{i=1}^{q} \beta_i E_i(t)[S_i(t) - \sigma_{ii}(E(t))] dt = \oint \sum_{i=1}^{q} \beta_i E_i(t) dQ_i(t) \geq 0, \quad (4.2)$$

where $E(t)$, $t \geq t_0$, is the solution to (3.5) with initial condition $E(t_0) = E_0$. Furthermore,

$$\oint \sum_{i=1}^{q} \frac{dQ_i(t)}{c + \beta_i E_i(t)} = 0 \qquad (4.3)$$

and

$$\oint \sum_{i=1}^{q} \beta_i E_i(t) dQ_i(t) = 0 \qquad (4.4)$$

if and only if there exists a continuous function $\alpha : [t_0, t_f] \to \overline{\mathbb{R}}_+$ such that $E(t) = \alpha(t)\boldsymbol{p}$, $t \in [t_0, t_f]$, where $\boldsymbol{p} \triangleq [1/\beta_1, ..., 1/\beta_q]^{\mathrm{T}}$.

Proof. The proof is identical to the proofs of Propositions 3.2 and 3.6. $\qquad\qquad\qquad\qquad\qquad\qquad\qquad\qquad\qquad\qquad\qquad\qquad\square$

Note that with the modified Axiom i) the isolated large-scale dynamical system $\mathcal{G}$ has equilibrium energy states given by $E_e = \alpha \boldsymbol{p}$ for $\alpha \geq 0$. As in Section 3.3, we define an equilibrium process as a process in which the trajectory of the system $\mathcal{G}$ moves along the equilibrium manifold $\mathcal{M}_e \triangleq \{E \in \overline{\mathbb{R}}_+^q : E = \alpha \boldsymbol{p}, \alpha \geq 0\}$ corresponding to the set of equilibria for the isolated system $\mathcal{G}$, and we define a nonequilibrium process as a process that does not lie on $\mathcal{M}_e$. Thus, it follows from Axioms i) and ii) that inequalities (4.1) and (4.2) are satisfied as equalities for an equilibrium process and as strict inequalities for a nonequilibrium process. Next, in light of our modified axioms, we present a generalized definition for the entropy and ectropy of $\mathcal{G}$.

Definition 4.1 *For the large-scale dynamical system $\mathcal{G}$ with power balance equation (3.5), a function $\mathcal{S} : \overline{\mathbb{R}}_+^q \to \mathbb{R}$ satisfying*

$$S(E(t_2)) \geq S(E(t_1)) + \int_{t_1}^{t_2} \sum_{i=1}^{q} \frac{S_i(t) - \sigma_{ii}(E(t))}{c + \beta_i E_i(t)} dt \qquad (4.5)$$

for any $t_2 \geq t_1 \geq t_0$ and $S(\cdot) \in \mathcal{U}$ is called the entropy *function of $\mathcal{G}$.*

Definition 4.2 *For the large-scale dynamical system $\mathcal{G}$ with power balance equation (3.5), a function $\mathcal{E} : \overline{\mathbb{R}}_+^q \to \mathbb{R}$ satisfying*

$$\mathcal{E}(E(t_2)) \leq \mathcal{E}(E(t_1)) + \int_{t_1}^{t_2} \sum_{i=1}^{q} \beta_i E_i(t)[S_i(t) - \sigma_{ii}(E(t))]dt \quad (4.6)$$

for any $t_2 \geq t_1 \geq t_0$ and $S(\cdot) \in \mathcal{U}$ is called the ectropy *function of $\mathcal{G}$.*

For the next result, define the available entropy and available ectropy of the large-scale dynamical system $\mathcal{G}$ by

$$\mathcal{S}_a(E_0) \triangleq - \sup_{S(\cdot) \in \mathcal{U}_c, T \geq t_0} \int_{t_0}^{T} \sum_{i=1}^{q} \frac{S_i(t) - \sigma_{ii}(E(t))}{c + \beta_i E_i(t)} dt, \qquad (4.7)$$

$$\mathcal{E}_a(E_0) \triangleq - \inf_{S(\cdot) \in \mathcal{U}_c, T \geq t_0} \int_{t_0}^{T} \sum_{i=1}^{q} \beta_i E_i(t)[S_i(t) - \sigma_{ii}(E(t))]dt, \quad (4.8)$$

where $E(t_0) = E_0 \in \overline{\mathbb{R}}_+^q$ and $E(T) = 0$, and define the required entropy supply and required ectropy supply of the large-scale dynamical system $\mathcal{G}$ by

$$\mathcal{S}_r(E_0) \triangleq \sup_{S(\cdot) \in \mathcal{U}_r, T \geq -t_0} \int_{-T}^{t_0} \sum_{i=1}^{q} \frac{S_i(t) - \sigma_{ii}(E(t))}{c + \beta_i E_i(t)} dt, \qquad (4.9)$$

$$\mathcal{E}_r(E_0) \triangleq \inf_{S(\cdot) \in \mathcal{U}_r, T \geq -t_0} \int_{-T}^{t_0} \sum_{i=1}^{q} \beta_i E_i(t)[S_i(t) - \sigma_{ii}(E(t))]dt, \quad (4.10)$$

where $E(-T) = 0$ and $E(t_0) = E_0 \in \overline{\mathbb{R}}_+^q$.

Theorem 4.1 *Consider the large-scale dynamical system $\mathcal{G}$ with power balance equation (3.5), and assume that Axiom ii) holds. Then there exist an entropy and an ectropy function for $\mathcal{G}$. Moreover,*

$\mathcal{S}_{\mathrm{a}}(E)$, $E \in \overline{\mathbb{R}}_+^q$, and $\mathcal{S}_{\mathrm{r}}(E)$, $E \in \overline{\mathbb{R}}_+^q$, are possible entropy functions for $\mathcal{G}$ with $\mathcal{S}_{\mathrm{a}}(0) = \mathcal{S}_{\mathrm{r}}(0) = 0$, and $\mathcal{E}_{\mathrm{a}}(E)$, $E \in \overline{\mathbb{R}}_+^q$, and $\mathcal{E}_{\mathrm{r}}(E)$, $E \in \overline{\mathbb{R}}_+^q$, are possible ectropy functions for $\mathcal{G}$ with $\mathcal{E}_{\mathrm{a}}(0) = \mathcal{E}_{\mathrm{r}}(0) = 0$. Finally, all entropy functions $\mathcal{S}(E)$, $E \in \overline{\mathbb{R}}_+^q$, for $\mathcal{G}$ satisfy

$$\mathcal{S}_{\mathrm{r}}(E) \le \mathcal{S}(E) - \mathcal{S}(0) \le \mathcal{S}_{\mathrm{a}}(E), \quad E \in \overline{\mathbb{R}}_+^q, \tag{4.11}$$

and all ectropy functions $\mathcal{E}(E)$, $E \in \overline{\mathbb{R}}_+^q$, for $\mathcal{G}$ satisfy

$$\mathcal{E}_{\mathrm{a}}(E) \le \mathcal{E}(E) - \mathcal{E}(0) \le \mathcal{E}_{\mathrm{r}}(E), \quad E \in \overline{\mathbb{R}}_+^q. \tag{4.12}$$

Proof. The proof is identical to the proofs of Theorems 3.2 and 3.6. □

The next series of results gives analogous results to the results in Sections 3.3 and 3.4 for the modified definitions of entropy and ectropy given in this chapter.

Theorem 4.2 *Consider the large-scale dynamical system $\mathcal{G}$ with power balance equation (3.5), and let $\mathcal{S} : \overline{\mathbb{R}}_+^q \to \mathbb{R}$ and $\mathcal{E} : \overline{\mathbb{R}}_+^q \to \mathbb{R}$ be entropy and ectropy functions of $\mathcal{G}$, respectively. Then $\mathcal{S}(\cdot)$ and $\mathcal{E}(\cdot)$ are continuous on $\overline{\mathbb{R}}_+^q$.*

Proof. The proof is identical to the proof of Theorem 3.3. □

For the statement of the next result, recall the definition of $\boldsymbol{p} = [1/\beta_1, ..., 1/\beta_q]^{\mathrm{T}}$ given in Proposition 4.1 and define $P \triangleq \mathrm{diag}[\beta_1, ..., \beta_q]$.

Proposition 4.2 *Consider the large-scale dynamical system $\mathcal{G}$ with power balance equation (3.5), and assume that Axioms i) and ii) hold. Then at every equilibrium state $E_{\mathrm{e}} = \alpha \boldsymbol{p}$, $\alpha \ge 0$, of the isolated system $\mathcal{G}$, the entropy $\mathcal{S}(E)$, $E \in \overline{\mathbb{R}}_+^q$, and ectropy $\mathcal{E}(E)$, $E \in \overline{\mathbb{R}}_+^q$, functions of $\mathcal{G}$ are unique (modulo a constant of integration) and are given by*

$$\mathcal{S}(E) - \mathcal{S}(0) = \mathcal{S}_{\mathrm{a}}(E) = \mathcal{S}_{\mathrm{r}}(E) = \boldsymbol{p}^{\mathrm{T}}\mathbf{log}_e(c\boldsymbol{e} + PE) - \boldsymbol{e}^{\mathrm{T}}\boldsymbol{p}\log_e c \tag{4.13}$$

and

$$\mathcal{E}(E) - \mathcal{E}(0) = \mathcal{E}_{\mathrm{a}}(E) = \mathcal{E}_{\mathrm{r}}(E) = \tfrac{1}{2}E^{\mathrm{T}}PE, \tag{4.14}$$

respectively, where $E = E_{\mathrm{e}}$ and $\mathbf{log}_e(c\boldsymbol{e} + PE)$ denotes the vector natural logarithm given by $[\log_e(c + \beta_1 E_1), ..., \log_e(c + \beta_q E_q)]^{\mathrm{T}}$.

Proof. The proof is identical to the proofs of Propositions 3.4 and 3.8. □

Proposition 4.3 *Consider the large-scale dynamical system $\mathcal{G}$ with power balance equation (3.5), and assume that Axioms i) and ii) hold. Let $\mathcal{S}(\cdot)$ and $\mathcal{E}(\cdot)$ denote an entropy and ectropy of $\mathcal{G}$, respectively, and let $E : [t_0, t_1] \to \overline{\mathbb{R}}_+^q$ denote the solution to (3.5) with $E(t_0) = \alpha_0 \boldsymbol{p}$ and $E(t_1) = \alpha_1 \boldsymbol{p}$, where $\alpha_0, \alpha_1 \geq 0$. Then*

$$\mathcal{S}(E(t_1)) = \mathcal{S}(E(t_0)) + \int_{t_0}^{t_1} \sum_{i=1}^{q} \frac{S_i(t) - \sigma_{ii}(E(t))}{c + \beta_i E_i(t)} dt \qquad (4.15)$$

and

$$\mathcal{E}(E(t_1)) = \mathcal{E}(E(t_0)) + \int_{t_0}^{t_1} \sum_{i=1}^{q} \beta_i E_i(t)[S_i(t) - \sigma_{ii}(E(t))]dt \qquad (4.16)$$

if and only if there exists a continuous function $\alpha : [t_0, t_1] \to \overline{\mathbb{R}}_+$ such that $\alpha(t_0) = \alpha_0$, $\alpha(t_1) = \alpha_1$, and $E(t) = \alpha(t)\boldsymbol{p}$, $t \in [t_0, t_1]$.

Proof. The proof is identical to the proofs of Propositions 3.5 and 3.9. $\qquad \square$

Theorem 4.3 *Consider the large-scale dynamical system $\mathcal{G}$ with power balance equation (3.5), and assume that Axioms i) and ii) hold. Then the function $\mathcal{S} : \overline{\mathbb{R}}_+^q \to \mathbb{R}$ given by*

$$\mathcal{S}(E) = \boldsymbol{p}^{\mathrm{T}}\log_e(c\boldsymbol{e} + PE) - \boldsymbol{e}^{\mathrm{T}}\boldsymbol{p}\log_e c, \quad E \in \overline{\mathbb{R}}_+^q, \qquad (4.17)$$

is a unique (modulo a constant of integration), continuously differentiable entropy function of $\mathcal{G}$. Furthermore, the function $\mathcal{E} : \overline{\mathbb{R}}_+^q \to \mathbb{R}$ given by

$$\mathcal{E}(E) = \tfrac{1}{2}E^{\mathrm{T}}PE, \quad E \in \overline{\mathbb{R}}_+^q, \qquad (4.18)$$

is a unique (modulo a constant of integration), continuously differentiable ectropy function of $\mathcal{G}$. In addition, for $E(t) \notin \mathcal{M}_e$, $t \geq t_0$, where $E(t)$, $t \geq t_0$, denotes the solution to (3.5) and $\mathcal{M}_e = \{E \in \overline{\mathbb{R}}_+^q : E = \alpha\boldsymbol{p}, \alpha \geq 0\}$, (4.17) and (4.18) satisfy

$$\mathcal{S}(E(t_2)) > \mathcal{S}(E(t_1)) + \int_{t_1}^{t_2} \sum_{i=1}^{q} \frac{S_i(t) - \sigma_{ii}(E(t))}{c + \beta_i E_i(t)} dt \qquad (4.19)$$

and

$$\mathcal{E}(E(t_2)) < \mathcal{E}(E(t_1)) + \int_{t_1}^{t_2} \sum_{i=1}^{q} \beta_i E_i(t)[S_i(t) - \sigma_{ii}(E(t))]dt \qquad (4.20)$$

for any $t_2 \geq t_1 \geq t_0$ and $\mathcal{S}(\cdot) \in \mathcal{U}$.

Proof. The proof is identical to the proofs of Theorems 3.4 and 3.8. $\square$

It is important to note that Theorem 4.3 establishes the existence of a unique entropy and ectropy function for $\mathcal{G}$ for equilibrium and nonequilibrium processes. Furthermore, it follows from Theorem 4.3 that the entropy and ectropy functions for $\mathcal{G}$ defined by (4.17) and (4.18) satisfy, respectively, (4.5) and (4.6) as equalities for an equilibrium process and as strict inequalities for a nonequilibrium process. Hence, it follows from Theorem 2.15 that the isolated large-scale dynamical system $\mathcal{G}$ does not exhibit Poincaré recurrence in $\overline{\mathbb{R}}_+^q \setminus \mathcal{M}_e$.

Once again, inequality (4.5) is a generalized Clausius inequality for equilibrium and nonequilibrium thermodynamics, while inequality (4.6) is an anti–Clausius inequality. Moreover, for the ectropy function given by (4.18), inequality (4.6) shows that a thermodynamically consistent large-scale dynamical system model is dissipative with respect to the supply rate $E^T P S$ and with storage function corresponding to the system ectropy $\mathcal{E}(E)$. In addition, if we let $dQ_i(t) = [S_i(t) - \sigma_{ii}(E(t))]dt$, $i = 1, ..., q$, denote the infinitesimal amount of the net heat received or dissipated by the ith subsystem of $\mathcal{G}$ over the infinitesimal time interval dt at the (shifted) *absolute ith subsystem temperature* $T_i \triangleq c + \beta_i E_i$, then it follows from (4.5) that the system entropy varies by an amount

$$dS(E(t)) \geq \sum_{i=1}^q \frac{dQ_i(t)}{c + \beta_i E_i(t)}, \quad t \geq t_0. \tag{4.21}$$

In light of the above definition of temperature, it is important to note that if $\beta_i \neq \beta_j$ for some $i \neq j$, then our thermodynamically consistent large-scale system model allows for the consideration of subsystems that possess the same stored energy, with one subsystem being hotter than the other.

Finally, note that the nonconservation of entropy and ectropy equations (3.105) and (3.106), respectively, for isolated large-scale dynamical systems also hold for the more general definitions of entropy and ectropy given in Definitions 4.1 and 4.2. In addition, using the modified definitions of entropy and ectropy given in Definitions 4.1 and 4.2 and using similar arguments as in Section 3.3, it can be shown that for every $E_0 \notin \mathcal{M}_e = \{E \in \overline{\mathbb{R}}_+^q : E = \alpha p, \alpha \geq 0\}$, the nonlinear dynamical system $\mathcal{G}$ with power balance equation (3.5) is state irreversible.

The following theorem is a generalization of Theorem 3.9.

Theorem 4.4 *Consider the large-scale dynamical system $\mathcal{G}$ with power balance equation (3.5) with $S(t) \equiv 0$ and $d(E) \equiv 0$, and assume that Axioms i) and ii) hold. Then for every $\alpha \geq 0$, $\alpha\boldsymbol{p}$ is a semistable equilibrium state of (3.5). Furthermore, $E(t) \to \frac{1}{\mathbf{e}^{\mathrm{T}}\boldsymbol{p}}\boldsymbol{p}\mathbf{e}^{\mathrm{T}}E(t_0)$ as $t \to \infty$ and $\frac{1}{\mathbf{e}^{\mathrm{T}}\boldsymbol{p}}\boldsymbol{p}\mathbf{e}^{\mathrm{T}}E(t_0)$ is a semistable equilibrium state. Finally, if for some $k \in \{1, ..., q\}$, $\sigma_{kk}(E) \geq 0$ and $\sigma_{kk}(E) = 0$ if and only if $E_k = 0$, then the zero solution $E(t) \equiv 0$ to (3.5) is a globally asymptotically stable equilibrium state of (3.5).*

Proof. It follows from Axiom i) that $\alpha\boldsymbol{p} \in \overline{\mathbb{R}}_+^q$, $\alpha \geq 0$, is an equilibrium state of (3.5). To show Lyapunov stability of the equilibrium state $\alpha\boldsymbol{p}$, consider the system-shifted ectropy $\mathcal{E}_{\mathrm{s}}(E) = \frac{1}{2}(E - \alpha\boldsymbol{p})^{\mathrm{T}}P(E - \alpha\boldsymbol{p})$ as a Lyapunov function candidate. Now, the proof follows as in the proof of Theorem 3.9 by invoking Axiom ii) and noting that $\phi_{ij}(E) = -\phi_{ji}(E)$, $E \in \overline{\mathbb{R}}_+^q$, $i \neq j$, $i, j = 1, ..., q$, $P\boldsymbol{p} = \mathbf{e}$, and $\mathbf{e}^{\mathrm{T}}w(E) = 0$, $E \in \overline{\mathbb{R}}_+^q$. Alternatively, in the case where for some $k \in \{1, ..., q\}$, $\sigma_{kk}(E) \geq 0$ and $\sigma_{kk}(E) = 0$ if and only if $E_k = 0$, global asymptotic stability of the zero solution $E(t) \equiv 0$ to (3.5) follows from standard Lyapunov arguments using the system ectropy $\mathcal{E}(E) = \frac{1}{2}E^{\mathrm{T}}PE$ as a Lyapunov function candidate. $\square$

It follows from Theorem 4.4 that the steady-state value of the energy in each subsystem $\mathcal{G}_i$ of the isolated large-scale dynamical system $\mathcal{G}$ is given by $E_\infty = \frac{1}{\mathbf{e}^{\mathrm{T}}\boldsymbol{p}}\boldsymbol{p}\mathbf{e}^{\mathrm{T}}E(t_0)$, which implies that $E_{i\infty} = \frac{1}{\beta_i\mathbf{e}^{\mathrm{T}}\boldsymbol{p}}\mathbf{e}^{\mathrm{T}}E(t_0)$ or, equivalently, $\hat{T}_{i\infty} = \beta_i E_{i\infty} = \frac{1}{\mathbf{e}^{\mathrm{T}}\boldsymbol{p}}\mathbf{e}^{\mathrm{T}}E(t_0)$. Hence, the steady-state temperature of the isolated large-scale dynamical system $\mathcal{G}$ given by $\hat{T}_\infty = \frac{1}{\mathbf{e}^{\mathrm{T}}\boldsymbol{p}}\mathbf{e}^{\mathrm{T}}E(t_0)\mathbf{e}$ is uniformly distributed over all the subsystems of $\mathcal{G}$. This phenomenon is known as *temperature equipartition*, in which all the system energy is eventually transformed into heat at a uniform temperature, and hence, all dynamic processes in $\mathcal{G}$ (system motions) would cease.

Proposition 4.4 *Consider the large-scale dynamical system $\mathcal{G}$ with power balance equation (3.5), let $\mathcal{E} : \overline{\mathbb{R}}_+^q \to \overline{\mathbb{R}}_+$ and $\mathcal{S} : \overline{\mathbb{R}}_+^q \to \overline{\mathbb{R}}_+$ denote the ectropy and entropy functions of $\mathcal{G}$ given by (4.18) and (4.17), respectively, and define $\mathcal{D}_{\mathrm{c}} \triangleq \{E \in \overline{\mathbb{R}}_+^q : \mathbf{e}^{\mathrm{T}}E = \beta\}$, where $\beta \geq 0$. Then*

$$\arg\min_{E \in \mathcal{D}_{\mathrm{c}}}(\mathcal{E}(E)) = \arg\max_{E \in \mathcal{D}_{\mathrm{c}}}(\mathcal{S}(E)) = E^* = \frac{\beta}{\mathbf{e}^{\mathrm{T}}\boldsymbol{p}}\boldsymbol{p}. \qquad (4.22)$$

Furthermore, $\mathcal{E}_{\min} \triangleq \mathcal{E}(E^*) = \frac{1}{2}\frac{\beta^2}{\mathbf{e}^{\mathrm{T}}\boldsymbol{p}}$ and $\mathcal{S}_{\max} \triangleq \mathcal{S}(E^*) = \mathbf{e}^{\mathrm{T}}\boldsymbol{p}\log_e(c+$

$\frac{\beta}{e^{T}\boldsymbol{p}}) - e^{T}\boldsymbol{p}\log_{e} c.$

Proof. The proof is identical to the proof of Proposition 3.10 and hence is omitted. $\square$

Proposition 4.4 shows that when all the energy of the large-scale dynamical system $\mathcal{G}$ is transformed into heat at a uniform temperature, the system entropy is a maximum and the system ectropy is a minimum.

4.2 Boltzmann Thermodynamics

As noted in Chapter 1, Boltzmann [15] was the first to give a probabilistic interpretation of entropy using a microscopic point of view of molecules in a system. In particular, probability was used in the context of a measure of the variety of ways in which the molecules in a system can be rearranged without changing the macroscopic properties of the system. Specifically, realizing that a system macrostate can be represented by many different microstates involving different configurations of molecular motion, macroscopic phenomena can be derived from the microscopic dynamics. Hence, the entropy of an observed macroscopic state is defined as the logarithmic probability of its occurrence. Since entropy measures probability and probability, in turn, expresses disorder, entropy is a measure of disorder. This is perhaps best reflected in Boltzmann's kinetic theory of gases, in which the entropy of a gas, defined in terms of the probability distribution, increases as a more uniformly distributed state is reached when the gas diffuses from a filled container into an empty container. Hence, the entropy of the gas increases until the system reaches the configuration with the largest number of microscopic states, the most probable configuration. Since the final system state can be realized in many more ways than the initial, more organized system state, it has the highest probability and hence the maximal entropy.

In this section, we provide a deterministic kinetic theory interpretation of the steady-state expressions for the entropy and ectropy presented in this chapter. Specifically, we assume that each subsystem $\mathcal{G}_i$ of the large-scale dynamical system $\mathcal{G}$ is a simple system consisting of an ideal gas with rigid walls. Furthermore, we assume that all subsystems $\mathcal{G}_i$ are divided by *diathermal walls* (that is, walls that permit energy flow) and the overall dynamical system is a closed system (that is, the system is separated from the environment by a rigid adiabatic wall). In this case, $\beta_i = k/n_i$, $i = 1, \ldots, q$, where n_i, $i = 1, \ldots, q$, is the number of molecules in the ith subsystem and $k > 0$ is the *Boltzmann constant*[2] (i.e., the gas constant per molecule). Without loss of generality and for simplicity of exposition let $k = 1$. In addition, we assume that the molecules in the ideal gas are hard elastic spheres, that is, there are no forces between the molecules except during collisions, and the molecules are not deformed by collisions. Thus, there is no internal potential energy, and the system internal energy of the ideal gas is entirely kinetic. Hence, in this case, the temperature of each subsystem $\mathcal{G}_i$ is the average translational kinetic energy per molecule, which is consistent with the kinetic theory of ideal gases.

Definition 4.3 *For a given isolated large-scale dynamical system $\mathcal{G}$ in thermal equilibrium, define the equilibrium entropy of $\mathcal{G}$ by $\mathcal{S}_e = n \log_e (c + \frac{\mathbf{e}^{\mathrm{T}} E_\infty}{n}) - n \log_e c$ and the equilibrium ectropy of $\mathcal{G}$ by $\mathcal{E}_e = \frac{1}{2} \frac{(\mathbf{e}^{\mathrm{T}} E_\infty)^2}{n}$, where $\mathbf{e}^{\mathrm{T}} E_\infty$ denotes the total steady-state energy of the large-scale dynamical system $\mathcal{G}$ and n denotes the total number of molecules in $\mathcal{G}$.*

Note that the definitions of equilibrium entropy and equilibrium ectropy given in Definition 4.3 are entirely consistent with the equilibrium (maximum) entropy and equilibrium (minimum) ectropy given by Proposition 4.4. Next, assume that each subsystem $\mathcal{G}_i$ is initially in thermal equilibrium. Furthermore, for each subsystem, let E_i and n_i, $i = 1, \ldots, q$, denote the total internal energy and the number of molecules, respectively, in the ith subsystem. Hence, the entropy and ectropy of the ith subsystem are given by $\mathcal{S}_i = n_i \log_e (c + E_i/n_i) - n_i \log_e c$ and $\mathcal{E}_i = \frac{1}{2} \frac{E_i^2}{n_i}$, respectively. Next, note that the entropy and the ectropy of the overall system (after reaching a thermal equilibrium) are given by $\mathcal{S}_e = n \log_e (c + \frac{\mathbf{e}^{\mathrm{T}} E_\infty}{n}) - n \log_e c$ and $\mathcal{E}_e = \frac{1}{2} \frac{(\mathbf{e}^{\mathrm{T}} E_\infty)^2}{n}$. Now, it follows from the convexity of $-\log_e(\cdot)$ and conservation of energy that the entropy of $\mathcal{G}$ at thermal equilibrium is given by

$$
\mathcal{S}_e = n \log_e \left(c + \frac{\mathbf{e}^{\mathrm{T}} E_\infty}{n} \right) - n \log_e c
$$

$$
= n \log_e \left[\sum_{i=1}^{q} \frac{n_i}{n} \left(c + \frac{E_i}{n_i} \right) \right] - \sum_{i=1}^{q} n_i \log_e c
$$

---

[2]The Boltzmann constant is equal to the ratio of the universal gas constant to Avogadro's number. This constant is physically significant but can be ignored from a mathematical perspective.

Figure 4.1 Entropy (respectively, ectropy) increases (respectively, decreases) as a
more evenly distributed state is reached.

$$\geq n \sum_{i=1}^{q} \frac{n_i}{n} \log_e \left(c + \frac{E_i}{n_i} \right) - \sum_{i=1}^{q} n_i \log_e c$$

$$= \sum_{i=1}^{q} \mathcal{S}_i. \tag{4.23}$$

Furthermore, the ectropy of $\mathcal{G}$ at thermal equilibrium is given by

$$\mathcal{E}_e = \frac{1}{2} \frac{(e^T E_\infty)^2}{n}$$

$$= \sum_{i=1}^{q} \frac{1}{2} \frac{E_i^2}{n_i} - \frac{1}{2n} \sum_{i=1}^{q} \sum_{j=i+1}^{q} \frac{(n_j E_i - n_i E_j)^2}{n_i n_j}$$

$$\leq \sum_{i=1}^{q} \frac{1}{2} \frac{E_i^2}{n_i}$$

$$= \sum_{i=1}^{q} \mathcal{E}_i. \tag{4.24}$$

It follows from (4.23) (respectively, (4.24)) that the equilibrium entropy (respectively, ectropy) of the system (gas) $\mathcal{G}$ is always greater
(respectively, less) than or equal to the sum of the entropies (respectively, ectropies) of the individual subsystems $\mathcal{G}_i$. Hence, the
entropy (respectively, ectropy) of the gas increases (respectively, decreases) as a more evenly distributed (disordered) state is reached
(see Figure 4.1). Finally, note that it follows from (4.23) and (4.24)
that $\mathcal{S}_e = \sum_{i=1}^{q} \mathcal{S}_i$ and $\mathcal{E}_e = \sum_{i=1}^{q} \mathcal{E}_i$ if and only if $\frac{E_i}{n_i} = \frac{E_j}{n_j}$, $i \neq$
j, $i, j = 1, ..., q$, that is, the initial temperatures of all subsystems are
equal. Furthermore, it follows from Axioms i) and ii) that the equal-
ity $\frac{E_i(t_0)}{n_i} = \frac{E_j(t_0)}{n_j}$, $i \neq j$, $i, j = 1, ..., q$, determines an equilibrium
state and hence a state reversible process (i.e., $\frac{E_i(t)}{n_i} = \frac{E_j(t)}{n_j}$, $t \geq t_0$,
$i \neq j$, $i, j = 1, ..., q$) for the system consisting of q ideal gases. In light
of the above, the following proposition is immediate.

Proposition 4.5 *For every state reversible adiabatic process performed on a system consisting of q ideal gases connected by diathermal walls, the total entropy and total ectropy of the system remain constant.*

Chapter Five

Work, Heat, and the Carnot Cycle

5.1 On the Equivalence of Work and Heat: The First Law Revisited

In Chapter 3, we showed that the first law of thermodynamics is essentially a statement of the principle of the conservation of energy. Hence, the variation in energy of a dynamical system $\mathcal{G}$ during any transformation is equal to the amount of energy that the system receives from the environment. In Chapter 3, however, the notion of energy that the system receives from the environment and dissipates to the environment was limited to heat and did not include *work*. When external forces act on the dynamical system $\mathcal{G}$, they can produce work on the system, changing the system's internal energy. Thus, addressing work performed by the system on the environment and work done by the environment on the system plays a crucial role in the principle of the conservation of energy for thermodynamic systems.

In this chapter we augment our nonlinear compartmental dynamical system model with an additional (deformation) state representing compartmental volumes in order to introduce the notion of work into our thermodynamically consistent energy flow model. Specifically, using Figure 5.1, we characterize a power balance equation such that during a dynamical transformation, the large-scale system $\mathcal{G}$ can perform (positive) work on its surroundings or the surroundings can do (negative) work on $\mathcal{G}$ resulting in subsystem volume changes. In this case, the power balance equation (3.5) takes the new form involving energy and deformation states given by

$$\dot{E}(t) = w(E(t), V(t)) - d_{\mathrm{w}}(E(t), V(t)) + S_{\mathrm{w}}(t) - d(E(t), V(t)) + S(t),$$
$$E(t_0) = E_0, \quad t \geq t_0, \quad (5.1)$$

$$\dot{V}_i(t) = \frac{[d_{\mathrm{w}i}(E(t), V(t)) - S_{\mathrm{w}i}(t)]V_i(t)}{(c + E_i(t))}, \quad V_i(t_0) = V_{i0}, \quad i = 1, ..., q,$$
$$(5.2)$$

where $c > 0$, $V(t) \triangleq [V_1(t), ..., V_q(t)]^{\mathrm{T}} \in \mathbb{R}_+^q$, $t \geq t_0$, $V_i : [0, \infty) \to \mathbb{R}_+$, $i = 1, ..., q$, denotes the volume of the ith subsystem, $V_{i0} > 0$, $i =$

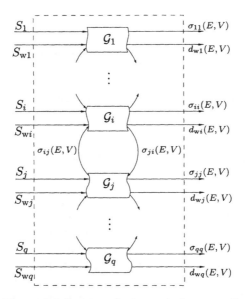

Figure 5.1 Large-scale dynamical system $\mathcal{G}$.

$1, ..., q$, $d_{\mathrm{w}}(E, V) = [d_{\mathrm{w}1}(E, V), ..., d_{\mathrm{w}q}(E, V)]^{\mathrm{T}}$, $d_{\mathrm{w}i} : \overline{\mathbb{R}}_+^q \times \mathbb{R}_+^q \rightarrow$ $\overline{\mathbb{R}}_+$, $i = 1, ..., q$, denotes the instantaneous rate of work done by the ith subsystem on the environment, $S_{\mathrm{w}}(t) = [S_{\mathrm{w}1}(t), ..., S_{\mathrm{w}q}(t)]^{\mathrm{T}}$, $t \geq t_0$, $S_{\mathrm{w}i} : [0, \infty) \rightarrow \overline{\mathbb{R}}_+$, $i = 1, ..., q$, denotes the instantaneous rate of work done by the environment on the ith subsystem, $d(E, V) = [\sigma_{11}(E, V), ..., \sigma_{qq}(E, V)]^{\mathrm{T}}$, $\sigma_{ii} : \overline{\mathbb{R}}_+^q \times \mathbb{R}_+^q \rightarrow \overline{\mathbb{R}}_+$, $i = 1, ..., q$, denotes the instantaneous rate of energy (heat) dissipation from the ith subsystem to the environment, $S(t) = [S_1(t), ..., S_q(t)]^{\mathrm{T}}$, $t \geq t_0$, $S_i : [0, \infty) \rightarrow \mathbb{R}$, $i = 1, ..., q$, denotes the external power (heat flux) supplied to (or extracted from) the ith subsystem, $w(E, V) \triangleq [w_1(E, V), ..., w_q(E, V)]^{\mathrm{T}}$, $w_i(E, V) = \sum_{j=1, j \neq i}^q [\sigma_{ij}(E, V) - \sigma_{ji}(E, V)]$, and $\sigma_{ij} :$ $\overline{\mathbb{R}}_+^q \times \mathbb{R}_+^q \rightarrow \overline{\mathbb{R}}_+$, $i \neq j$, $i, j = 1, ..., q$, denotes the instantaneous rate of energy (heat) flow from the jth subsystem to the ith subsystem.

As in Chapter 3, we assume that $\sigma_{ij}(E, V) = 0$, $E \in \overline{\mathbb{R}}_+^q$, $V \in \mathbb{R}_+^q$, whenever $E_j = 0$, $i, j = 1, ..., q$, and $S_i(t) \geq 0$ whenever $E_i(t) = 0$, $t \geq t_0$, $i = 1, ..., q$. Moreover, we assume that $d_{\mathrm{w}i}(E, V) = 0$, $E \in \overline{\mathbb{R}}_+^q$, $V \in \mathbb{R}_+^q$, whenever $E_i = 0$, $i = 1, ..., q$, which implies that if the energy of the ith subsystem is zero, then this subsystem cannot perform work on the environment. Finally, we assume that $\sigma_{ij} :$ $\overline{\mathbb{R}}_+^q \times \mathbb{R}_+^q \rightarrow \overline{\mathbb{R}}_+$, $i, j = 1, ..., q$, and $d_{\mathrm{w}i} : \overline{\mathbb{R}}_+^q \times \mathbb{R}_+^q \rightarrow \overline{\mathbb{R}}_+$, $i = 1, ..., q$, are locally Lipschitz continuous on $\overline{\mathbb{R}}_+^q \times \mathbb{R}_+^q$ and $S_i : [0, \infty) \rightarrow \mathbb{R}$, $i = 1, ..., q$, and $S_{\mathrm{w}i} : [0, \infty) \rightarrow \overline{\mathbb{R}}_+$, $i = 1, ..., q$, are piecewise continuous and bounded over the semi-infinite interval $[0, \infty)$. The above assumptions guarantee that the solution $[E^T(t), V^T(t)]^T$, $t \geq t_0$, to (5.1) and (5.2) exists and is nonnegative for all nonnegative initial conditions. Finally, note that (5.2) can be written in vector form as

$$\dot{V}(t) = D_p(E(t), V(t))[d_w(E(t), V(t)) - S_w(t)], \quad V(t_0) = V_0, \quad (5.3)$$

where $V_0 \in \mathbb{R}_+^q$ and $D_p(E, V) \triangleq \text{diag}[\frac{V_1}{c+E_1}, ..., \frac{V_q}{c+E_q}]$, $E \in \mathbb{R}_+^q$, $V \in \mathbb{R}_+^q$.

It follows from (5.1) and (5.2) that positive work done by a subsystem on the environment leads to a decrease in the internal energy of the subsystem and an increase in the subsystem volume, which is consistent with the first law of thermodynamics. To see that (5.1) and (5.2) is a statement of the first law of thermodynamics, define the work L done by the large-scale dynamical system $\mathcal{G}$ over the time interval $[t_1, t_2]$ by

$$L \triangleq \int_{t_1}^{t_2} \mathbf{e}^T [d_w(E(t), V(t)) - S_w(t)] dt, \quad (5.4)$$

where $[E^T(t), V^T(t)]^T$, $t \geq t_0$, is the solution to (5.1) and (5.2). Then, premultiplying (5.1) by $\mathbf{e}^T$ and using the fact that $\mathbf{e}^T w(E, V) \equiv 0$, it follows that

$$\Delta U = -L + Q, \quad (5.5)$$

where $\Delta U = U(t_2) - U(t_1) \triangleq \mathbf{e}^T E(t_2) - \mathbf{e}^T E(t_1)$ denotes the variation in total energy of the large-scale system $\mathcal{G}$ over the time interval $[t_1, t_2]$ and

$$Q \triangleq \int_{t_1}^{t_2} \mathbf{e}^T [S(t) - d(E(t), V(t))] dt \quad (5.6)$$

denotes the net energy received by $\mathcal{G}$ in forms other than work. This is a statement of the *first law of thermodynamics* for the large-scale dynamical system $\mathcal{G}$ and gives a precise formulation of the equivalence between work and heat. This establishes that heat and mechanical work are two different aspects of energy. For a cyclic transformation, the initial and final states of $\mathcal{G}$ are the same, and hence, the variation in energy is zero, that is, $\Delta U = 0$. Thus, (5.5) becomes

$$L = Q, \quad (5.7)$$

which shows that the work performed by the system over a cyclic transformation is equal to the net difference of the heat absorbed and the heat surrendered by the system. Finally, note that (5.2) is consistent with the classical thermodynamic equation for the rate of work done by the system on the environment with $\frac{c+E_i}{V_i}$ playing the role of subsystem *pressures*. To see this, note that (5.2) can be equivalently written as $dL_i = \left(\frac{c+E_i}{V_i}\right) dV_i$, which, for a single subsystem with volume V and pressure p, has the classical form

$$dL = pdV. \tag{5.8}$$

If the total energy of the large-scale dynamical system $\mathcal{G}$ at the initial and the final states is fixed, then it follows from (5.5) that the variation δ of the difference between the work done by the large-scale dynamical system $\mathcal{G}$ on the environment and the energy supplied to the large-scale dynamical system $\mathcal{G}$ satisfies

$$\delta(L - Q) = 0. \tag{5.9}$$

Equation (5.9) implies that if during a transformation between two fixed points the large-scale dynamical system $\mathcal{G}$ receives a fixed amount of energy, then the amount of work that the large-scale dynamical system can perform on the environment is also fixed. In other words, for any two paths connecting the initial and final states of the dynamical system $\mathcal{G}$ corresponding to identical energy supplies, the work done by the system is the same. In the case of an adiabatically isolated (i.e., $S(t) \equiv 0$ and $d(E, V) \equiv 0$) dynamical system $\mathcal{G}$, $Q = 0$, and hence, it follows from (5.9) that

$$\delta L = 0. \tag{5.10}$$

This implies that among the set of all possible smooth paths that an adiabatically isolated large-scale dynamical system $\mathcal{G}$ may move between two fixed points over a specified time interval, the only dynamically possible system paths are those that render the work L done by the system on the environment stationary to all variations in the shape of the paths. This is analogous to *Hamilton's principle of least action* in classical mechanics.

To guarantee a thermodynamically consistent energy flow model, we assume that Axioms *i*) and *ii*) given in Section 3.3 hold for the large-scale dynamical system $\mathcal{G}$ with $\phi_{ij}(E)$ replaced by $\phi_{ij}(E, V)$, $i \neq j$, $i, j = 1, ..., q$. In this case, the results of Section 3.3 pertaining to entropy also hold for the nonlinear thermodynamic model given by (5.1) and (5.2). However, the input spaces $\mathcal{U}$, $\mathcal{U}_r$, and $\mathcal{U}_c$ consist of subsets of bounded continuous $\mathbb{R}^q \times \mathbb{R}^q$-valued functions on $\mathbb{R}$. Furthermore, all the results regarding equilibrium and nonequilibrium processes also hold. Note that it follows from Axiom i) that for the isolated large-scale dynamical system $\mathcal{G}$ given by (5.1) and (5.2) (that is, $S(t) \equiv 0$, $d(E, V) \equiv 0$, $S_w(t) \equiv 0$, and $d_w(E, V) \equiv 0$), the points $(\alpha \mathbf{e}, V_e)$, where $\alpha \geq 0$ and V_e is an arbitrary point in $\mathbb{R}^q_+$, are the equilibrium states of $\mathcal{G}$. Here, we highlight the fundamental extensions corresponding to this generalization. For the next result, $\oint$ denotes a cyclic integral evaluated along an arbitrary closed path of (5.1) and (5.2) in $\overline{\mathbb{R}}^q_+ \times \mathbb{R}^q_+$; that is, $\oint \triangleq \int_{t_0}^{t_f}$ with $t_f \geq t_0$ and $(S(\cdot), S_w(\cdot)) \in \mathcal{U}$ such that $E(t_f) = E(t_0) = E_0 \in \overline{\mathbb{R}}^q_+$ and $V(t_f) = V(t_0) = V_0 \in \mathbb{R}^q_+$.

Proposition 5.1 *Consider the large-scale dynamical system $\mathcal{G}$ with power balance equations given by (5.1) and (5.2), and assume that Axioms i) and ii) hold. Then for all $E_0 \in \overline{\mathbb{R}}^q_+$, $V_0 \in \mathbb{R}^q_+$, $t_f \geq t_0$, $(S(\cdot), S_w(\cdot)) \in \mathcal{U}$ such that $E(t_f) = E(t_0) = E_0$ and $V(t_f) = V(t_0) = V_0$,*

$$\int_{t_0}^{t_f} \sum_{i=1}^{q} \frac{S_i(t) - \sigma_{ii}(E(t), V(t))}{c + E_i(t)} dt = \oint \sum_{i=1}^{q} \frac{dQ_i(t)}{c + E_i(t)} \leq 0, \quad (5.11)$$

where $c > 0$, $dQ_i(t) \triangleq [S_i(t) - \sigma_{ii}(E(t), V(t))]dt$, $i = 1, ..., q$, is the amount of the net energy (heat) received by the ith subsystem over the infinitesimal time interval dt and $[E^T(t), V^T(t)]^T$, $t \geq t_0$, is the solution to (5.1) and (5.2) with initial condition $[E^T(t_0), V^T(t_0)]^T = [E_0^T, V_0^T]^T$. Furthermore,

$$\oint \sum_{i=1}^{q} \frac{dQ_i(t)}{c + E_i(t)} = 0 \qquad (5.12)$$

if and only if there exists a continuous function $\alpha : [t_0, t_f] \to \overline{\mathbb{R}}_+$ such that $E(t) = \alpha(t)\mathbf{e}$ and $V(t) \in \mathbb{R}^q_+$, $t \in [t_0, t_f]$.

Proof. Since the solution to (5.1) and (5.2) is nonnegative and $\phi_{ij}(E, V) = -\phi_{ji}(E, V)$, $E \in \overline{\mathbb{R}}^q_+$, $V \in \mathbb{R}^q_+$, $i \neq j$, $i, j = 1, ..., q$, it follows from (5.1), (5.2), and Axiom ii) that

$$\oint \sum_{i=1}^{q} \frac{dQ_i(t)}{c + E_i(t)} = \int_{t_0}^{t_f} \sum_{i=1}^{q} \frac{\dot{E}_i(t) - \sum_{j=1, j \neq i}^{q} \phi_{ij}(E(t), V(t))}{c + E_i(t)} dt$$

$$+ \int_{t_0}^{t_f} \sum_{i=1}^{q} \frac{d_{wi}(E(t), V(t)) - S_{wi}(t)}{c + E_i(t)} dt$$

$$= \int_{t_0}^{t_f} \sum_{i=1}^{q} \frac{\dot{E}_i(t)}{c + E_i(t)} dt + \int_{t_0}^{t_f} \sum_{i=1}^{q} \frac{\dot{V}_i(t)}{V_i(t)} dt$$

$$-\int_{t_0}^{t_f} \sum_{i=1}^{q} \sum_{j=1, j\neq i}^{q} \frac{\phi_{ij}(E(t), V(t))}{c + E_i(t)} dt$$

$$= \sum_{i=1}^{q} \log_e \left(\frac{c + E_i(t_f)}{c + E_i(t_0)} \right) + \sum_{i=1}^{q} \log_e \left(\frac{V_i(t_f)}{V_i(t_0)} \right)$$

$$-\int_{t_0}^{t_f} \sum_{i=1}^{q} \sum_{j=i+1}^{q} \left(\frac{\phi_{ij}(E(t), V(t))}{c + E_i(t)} \right.$$

$$\left. - \frac{\phi_{ij}(E(t), V(t))}{c + E_j(t)} \right) dt$$

$$= -\int_{t_0}^{t_f} \sum_{i=1}^{q} \sum_{j=i+1}^{q} \frac{\phi_{ij}(E(t), V(t))(E_j(t) - E_i(t))}{(c + E_i(t))(c + E_j(t))} dt$$

$$\leq 0, \tag{5.13}$$

which proves (5.11).

To show (5.12), note that it follows from (5.13), Axiom i), and Axiom ii) that (5.12) holds if and only if $E_i(t) = E_j(t)$, $t \in [t_0, t_f]$, $i \neq j$, $i, j = 1, ..., q$, or, equivalently, there exists a continuous function $\alpha : [t_0, t_f] \to \overline{\mathbb{R}}_+$ such that $E(t) = \alpha(t)\mathbf{e}$ and $V(t) \in \mathbb{R}_+^q$, $t \in [t_0, t_f]$. $\square$

Next, we give a definition of entropy for the large-scale dynamical system $\mathcal{G}$ given by (5.1) and (5.2), which is consistent with the one given in Definition 3.2.

Definition 5.1 *For the large-scale dynamical system $\mathcal{G}$ with power balance equations given by (5.1) and (5.2), a function $\mathcal{S} : \overline{\mathbb{R}}_+^q \times \overline{\mathbb{R}}_+^q \to \mathbb{R}$ satisfying*

$$\mathcal{S}(E(t_2), V(t_2)) \geq \mathcal{S}(E(t_1), V(t_1))$$

$$+ \int_{t_1}^{t_2} \sum_{i=1}^{q} \frac{S_i(t) - \sigma_{ii}(E(t), V(t))}{c + E_i(t)} dt \tag{5.14}$$

for any $t_2 \geq t_1 \geq t_0$ and $(S(\cdot), S_{\mathrm{w}}(\cdot)) \in \mathcal{U}$ is called the entropy func-tion *of $\mathcal{G}$.*

As in Section 3.3, (5.11) guarantees the existence of an entropy function for $\mathcal{G}$. For the next result, define the available entropy of the large-scale dynamical system $\mathcal{G}$ by

$$\mathcal{S}_{\mathrm{a}}(E_0, V_0) \triangleq - \sup_{(S(\cdot), S_{\mathrm{w}}(\cdot)) \in \mathcal{U}_{\mathrm{c}}, T \geq t_0} \int_{t_0}^{T} \sum_{i=1}^{q} \frac{S_i(t) - \sigma_{ii}(E(t), V(t))}{c + E_i(t)} dt,$$

$$(5.15)$$

where $E(t_0) = E_0 \in \overline{\mathbb{R}}_+^q$, $V(t_0) = V_0 \in \mathbb{R}_+^q$, $E(T) = 0$, and $V(T) = V^*$, where $V^* \in \mathbb{R}_+^q$ denotes an arbitrary volume of $\mathcal{G}$ corresponding to the point of minimum system energy, and define the required entropy supply of the large-scale dynamical system $\mathcal{G}$ by

$$\mathcal{S}_{\mathrm{r}}(E_0, V_0) \triangleq \sup_{(S(\cdot), S_{\mathrm{w}}(\cdot)) \in \mathcal{U}_{\mathrm{r}}, \, T \geq -t_0} \int_{-T}^{t_0} \sum_{i=1}^{q} \frac{S_i(t) - \sigma_{ii}(E(t), V(t))}{c + E_i(t)} \, dt,$$

$$(5.16)$$

where $E(-T) = 0$, $V(-T) = V^*$, $E(t_0) = E_0 \in \overline{\mathbb{R}}_+^q$, and $V(t_0) = V_0 \in \mathbb{R}_+^q$.

Theorem 5.1 *Consider the large-scale dynamical system $\mathcal{G}$ with power balance equations given by (5.1) and (5.2), and assume that Axiom ii) holds. Then there exists an entropy function for $\mathcal{G}$. Moreover, $\mathcal{S}_{\mathrm{a}}(E, V)$, $(E, V) \in \overline{\mathbb{R}}_+^q \times \mathbb{R}_+^q$, and $\mathcal{S}_{\mathrm{r}}(E, V)$, $(E, V) \in \overline{\mathbb{R}}_+^q \times \mathbb{R}_+^q$, are possible entropy functions for $\mathcal{G}$ with $\mathcal{S}_{\mathrm{a}}(0, V^*) = \mathcal{S}_{\mathrm{r}}(0, V^*) = 0$. Finally, all entropy functions $\mathcal{S}(E, V)$, $(E, V) \in \overline{\mathbb{R}}_+^q \times \mathbb{R}_+^q$, for $\mathcal{G}$ satisfy*

$$\mathcal{S}_{\mathrm{r}}(E, V) \leq \mathcal{S}(E, V) - \mathcal{S}(0, V^*) \leq \mathcal{S}_{\mathrm{a}}(E, V), \quad (E, V) \in \overline{\mathbb{R}}_+^q \times \mathbb{R}_+^q.$$

$$(5.17)$$

Proof. The proof is similar to the proof of Theorem 3.2.   □

The following theorem shows that all entropy functions for $\mathcal{G}$ are continuous on $\overline{\mathbb{R}}_+^q \times \mathbb{R}_+^q$.

Theorem 5.2 *Consider the large-scale dynamical system $\mathcal{G}$ with power balance equations given by (5.1) and (5.2), and let $\mathcal{S} : \overline{\mathbb{R}}_+^q \times \mathbb{R}_+^q \to \mathbb{R}$ be an entropy function of $\mathcal{G}$. Then $\mathcal{S}(\cdot, \cdot)$ is continuous on $\overline{\mathbb{R}}_+^q \times \mathbb{R}_+^q$.*

Proof. Let $E_{\mathrm{e}} \in \overline{\mathbb{R}}_+^q$, $V_{\mathrm{e}} \in \mathbb{R}_+^q$, $S_{\mathrm{we}} \in \mathbb{R}^q$, and $S_{\mathrm{e}} \in \mathbb{R}^q$ be such that $S_{\mathrm{we}i} = d_{\mathrm{w}i}(E_{\mathrm{e}}, V_{\mathrm{e}})$ and $S_{\mathrm{e}i} = \sigma_{ii}(E_{\mathrm{e}}, V_{\mathrm{e}}) - \sum_{j=1, j \neq i}^{q} \phi_{ij}(E_{\mathrm{e}}, V_{\mathrm{e}})$, $i = 1, \ldots, q$. Note that with $S(t) \equiv S_{\mathrm{e}}$ and $S_{\mathrm{w}}(t) \equiv S_{\mathrm{we}}$, $[E_{\mathrm{e}}^{\mathrm{T}}, V_{\mathrm{e}}^{\mathrm{T}}]^{\mathrm{T}}$ is an equilibrium point of the dynamical system (5.1) and (5.2). Next, define $x \triangleq [E^{\mathrm{T}}, V^{\mathrm{T}}]^{\mathrm{T}}$ and $u \triangleq [S^{\mathrm{T}}, S_{\mathrm{w}}^{\mathrm{T}}]^{\mathrm{T}}$, and consider the linearization of (5.1) and (5.2) at $x_{\mathrm{e}} = [E_{\mathrm{e}}^{\mathrm{T}}, V_{\mathrm{e}}^{\mathrm{T}}]^{\mathrm{T}}$ and $u_{\mathrm{e}} = [S_{\mathrm{e}}^{\mathrm{T}}, S_{\mathrm{we}}^{\mathrm{T}}]^{\mathrm{T}}$ given by

$$\dot{x}(t) = A(x(t) - x_{\mathrm{e}}) + B(u(t) - u_{\mathrm{e}}), \quad x(t_0) = x_0, \quad t \geq t_0, \quad (5.18)$$

where

$$A = \left. \frac{\partial f(x)}{\partial x} \right|_{x=x_e}, \tag{5.19}$$

$$f(x) \triangleq [(w(x) - d_w(x) - d(x))^T, (D_p(x)d_w(x))^T]^T, \tag{5.20}$$

$$B = \begin{bmatrix} I_q & I_q \\ 0 & -D_p(x_e) \end{bmatrix}. \tag{5.21}$$

Since rank $B = 2q$ for all $x_e \in \overline{\mathbb{R}}_+^q \times \mathbb{R}_+^q$, it follows that

$$\text{rank}\,[B, AB, A^2B, ..., A^{2q-1}B] = 2q, \tag{5.22}$$

and hence, the linearized system (5.18) is controllable. The remainder of the proof now follows identically as in the proof of Theorem 3.3 using a slight generalization of Lemma 3.1.   □

The next result is a direct consequence of Theorem 5.1.

Proposition 5.2 *Consider the large-scale dynamical system* $\mathcal{G}$ *with power balance equations given by (5.1) and (5.2), and assume that Axioms i) and ii) hold. Then*

$$\mathcal{S}(E, V) \triangleq \alpha \mathcal{S}_r(E, V) + (1 - \alpha)\mathcal{S}_a(E, V), \quad \alpha \in [0, 1], \tag{5.23}$$

is an entropy function for $\mathcal{G}$.

The following propositions address equilibrium processes of $\mathcal{G}$ with the power balance equations (5.1) and (5.2).

Proposition 5.3 *Consider the large-scale dynamical system* $\mathcal{G}$ *with power balance equations given by (5.1) and (5.2), and assume that Axioms i) and ii) hold. Then at every equilibrium state* $(E_e, V_e) = (\alpha e, V_e)$, *where* $\alpha \geq 0$ *and* $V_e \in \mathbb{R}_+^q$, *of the isolated system* $\mathcal{G}$, *the entropy* $\mathcal{S}(E, V)$, $(E, V) \in \overline{\mathbb{R}}_+^q \times \mathbb{R}_+^q$, *of* $\mathcal{G}$ *is unique (modulo a constant of integration) and is given by*

$$\begin{aligned} \mathcal{S}(E, V) - \mathcal{S}(0, V^*) &= \mathcal{S}_a(E, V) \\ &= \mathcal{S}_r(E, V) \\ &= \mathbf{e}^T \log_e(ce + E) + \mathbf{e}^T \log_e V - \mathbf{e}^T \log_e V^* \\ &\quad - q \log_e c, \end{aligned} \tag{5.24}$$

where $E = E_e$ *and* $V = V_e$.

Proof. The proof is identical to the proof of Proposition 3.4.   □

Proposition 5.4 *Consider the large-scale dynamical system $\mathcal{G}$ with power balance equations given by (5.1) and (5.2), and assume that Axioms i) and ii) hold. Let $\mathcal{S}(\cdot,\cdot)$ denote an entropy of $\mathcal{G}$, and let $(E(t), V(t))$, $t \geq t_0$, denote the solution to (5.1) and (5.2) with $E(t_0) = \alpha_0\mathbf{e}$, $E(t_1) = \alpha_1\mathbf{e}$, and $V(t_0) = V_0 \in \mathbb{R}_+^q$, where α_0, $\alpha_1 \geq 0$. Then*

$$\mathcal{S}(E(t_1), V(t_1)) = \mathcal{S}(E(t_0), V(t_0))$$

$$+ \int_{t_0}^{t_1} \sum_{i=1}^{q} \frac{S_i(t) - \sigma_{ii}(E(t), V(t))}{c + E_i(t)} dt \qquad (5.25)$$

if and only if there exists a continuous function $\alpha : [t_0, t_1] \to \overline{\mathbb{R}}_+$ such that $\alpha(t_0) = \alpha_0$, $\alpha(t_1) = \alpha_1$, $E(t) = \alpha(t)\mathbf{e}$, and $V(t) \in \mathbb{R}_+^q$, $t \in [t_0, t_1]$.

Proof. The proof is identical to the proof of Proposition 3.5. $\square$

The next result gives a unique, continuously differentiable entropy function for $\mathcal{G}$ for equilibrium and nonequilibrium processes.

Theorem 5.3 *Consider the large-scale dynamical system $\mathcal{G}$ with power balance equations given by (5.1) and (5.2), and assume that Axioms i) and ii) hold. Then the function $\mathcal{S} : \overline{\mathbb{R}}_+^q \times \mathbb{R}_+^q \to \mathbb{R}$ given by*

$$\mathcal{S}(E, V) = \mathbf{e}^{\mathrm{T}}\log_e(c\mathbf{e} + E) + \mathbf{e}^{\mathrm{T}}\log_e V - \mathbf{e}^{\mathrm{T}}\log_e V^* - q\log_e c,$$
$$(E, V) \in \overline{\mathbb{R}}_+^q \times \mathbb{R}_+^q, \qquad (5.26)$$

where $c > 0$ and $V^* \in \mathbb{R}_+^q$, is a unique (modulo a constant of integration), continuously differentiable entropy function of $\mathcal{G}$. Furthermore, for $(E(t), V(t)) \notin \mathcal{M}_e$, $t \geq t_0$, where $(E(t), V(t))$, $t \geq t_0$, denotes the solution to (5.1), (5.2) and $\mathcal{M}_e = \{(E, V) \in \overline{\mathbb{R}}_+^q \times \mathbb{R}_+^q : E = \alpha\mathbf{e}, \alpha \geq 0, V \in \mathbb{R}_+^q\}$, (5.26) satisfies

$$\mathcal{S}(E(t_2), V(t_2)) > \mathcal{S}(E(t_1), V(t_1))$$

$$+ \int_{t_1}^{t_2} \sum_{i=1}^{q} \frac{S_i(t) - \sigma_{ii}(E(t), V(t))}{c + E_i(t)} dt \qquad (5.27)$$

for any $t_2 \geq t_1 \geq t_0$ and $(S(\cdot), S_{\mathrm{w}}(\cdot)) \in \mathcal{U}$.

Proof. Since the solution $[E^{\mathrm{T}}(t), V^{\mathrm{T}}(t)]^{\mathrm{T}}$, $t \geq t_0$, to (5.1) and (5.2) is nonnegative for all nonnegative initial conditions and $\phi_{ij}(E, V) = -\phi_{ji}(E, V)$, $E \in \overline{\mathbb{R}}_+^q$, $V \in \mathbb{R}_+^q$, $i \neq j$, $i, j = 1, ..., q$, it follows that

$$\dot{\mathcal{S}}(E(t), V(t)) = \sum_{i=1}^{q} \frac{\dot{E}_i(t)}{c + E_i(t)} + \sum_{i=1}^{q} \frac{\dot{V}_i(t)}{V_i(t)}$$

$$= \sum_{i=1}^{q} \frac{S_i(t) - \sigma_{ii}(E(t), V(t))}{c + E_i(t)}$$

$$+ \sum_{i=1}^{q} \frac{S_{\mathrm{w}i}(t) - d_{\mathrm{w}i}(E(t), V(t))}{c + E_i(t)}$$

$$+ \sum_{i=1}^{q} \sum_{j=1, j \neq i}^{q} \frac{\phi_{ij}(E(t), V(t))}{c + E_i(t)}$$

$$+ \sum_{i=1}^{q} \frac{d_{\mathrm{w}i}(E(t), V(t)) - S_{\mathrm{w}i}(t)}{c + E_i(t)}$$

$$= \sum_{i=1}^{q} \frac{S_i(t) - \sigma_{ii}(E(t), V(t))}{c + E_i(t)}$$

$$+ \sum_{i=1}^{q} \sum_{j=i+1}^{q} \left(\frac{\phi_{ij}(E(t), V(t))}{c + E_i(t)} - \frac{\phi_{ij}(E(t), V(t))}{c + E_j(t)} \right)$$

$$= \sum_{i=1}^{q} \frac{S_i(t) - \sigma_{ii}(E(t), V(t))}{c + E_i(t)}$$

$$+ \sum_{i=1}^{q} \sum_{j=i+1}^{q} \frac{\phi_{ij}(E(t), V(t))(E_j(t) - E_i(t))}{(c + E_i(t))(c + E_j(t))}$$

$$\geq \sum_{i=1}^{q} \frac{S_i(t) - \sigma_{ii}(E(t), V(t))}{c + E_i(t)}, \quad t \geq t_0. \tag{5.28}$$

Now, integrating (5.28) over $[t_1, t_2]$ yields (5.14). Furthermore, in the case where $(E(t), V(t)) \notin \mathcal{M}_{\mathrm{e}}$, $t \geq t_0$, it follows from Axiom i), Axiom ii), and (5.28) that (5.27) holds.

Uniqueness (modulo a constant of integration) of (5.26) follows using identical arguments as in the proof of Theorem 3.4. $\qquad \square$

Note that for an adiabatically isolated large-scale dynamical system $\mathcal{G}$, (5.14) yields the inequality

$$\mathcal{S}(E(t_2), V(t_2)) \geq \mathcal{S}(E(t_1), V(t_1)), \quad t_2 \geq t_1. \tag{5.29}$$

In addition, for the entropy function given by (5.26), inequality (5.29) is satisfied as a strict inequality for all $(E, V) \in \overline{\mathbb{R}}_+^q \times \mathbb{R}_+^q \setminus \mathcal{M}_{\mathrm{e}}$. Hence, it follows from Theorem 2.15 that the adiabatically isolated large-scale dynamical system $\mathcal{G}$ does not exhibit Poincaré recurrence in $\overline{\mathbb{R}}_+^q \times \mathbb{R}_+^q \setminus \mathcal{M}_{\mathrm{e}}$. Furthermore, using a similar analysis as given in the proof of Proposition 3.10, it can be shown that the entropy function given by (5.26) achieves its maximum among all the states of $\mathcal{G}$ with the fixed total energy and the fixed total volume of the system at $E = \alpha \mathbf{e}$ and $V = \gamma \mathbf{e}$, $\alpha \geq 0$, $\gamma > 0$. Hence, the maximum system entropy is attained when the energies and volumes of all subsystems of $\mathcal{G}$ are equal. Finally, for the entropy function given by (5.26), note that $\mathcal{S}(0, V^*) = 0$ or, equivalently, $\lim_{E \to 0} \mathcal{S}(E, V^*) = \mathcal{S}(0, V^*) = 0$, which is consistent with the third law of thermodynamics.

The following result is a slight extension to Theorem 3.5 and shows that the dynamical system $\mathcal{G}$ with power balance equations (5.1) and (5.2) is state irreversible for every nontrivial trajectory of $\mathcal{G}$. For this result, define the augmented state $x \triangleq [E^{\mathrm{T}}, V^{\mathrm{T}}]^{\mathrm{T}}$ and the augmented input $u \triangleq [S^{\mathrm{T}}, S_{\mathrm{w}}^{\mathrm{T}}]^{\mathrm{T}}$, and let $\mathcal{W}_{[t_0, t_1]}$ denote the set of all possible energy and volume trajectories of $\mathcal{G}$ over the time interval $[t_0, t_1]$ given by

$$\mathcal{W}_{[t_0,t_1]} \triangleq \{s^x : [t_0, t_1] \times \mathcal{U} \to \overline{\mathbb{R}}_+^q \times \mathbb{R}_+^q :$$
$$s^x(\cdot, u(\cdot)) \text{ satisfies (5.1) and (5.2)}\}. \quad (5.30)$$

Furthermore, let $\mathcal{M}_{\mathrm{e}} \subset \overline{\mathbb{R}}_+^q \times \mathbb{R}_+^q$ denote the set of equilibria of $\mathcal{G}$ given by $\mathcal{M}_{\mathrm{e}} = \{(E, V) \in \overline{\mathbb{R}}_+^q \times \mathbb{R}_+^q : E = \alpha \mathbf{e}, \alpha \geq 0, V \in \mathbb{R}_+^q\}$.

Theorem 5.4 *Consider the large-scale dynamical system $\mathcal{G}$ with power balance equations given by (5.1) and (5.2), and assume that Axioms i) and ii) hold. Furthermore, let $s^x(\cdot, u(\cdot)) \in \mathcal{W}_{[t_0,t_1]}$, where $u(\cdot) \in \mathcal{U}$. Then $s^x(\cdot, u(\cdot))$ is an I_{2q}-reversible trajectory of $\mathcal{G}$ if and only if $s^x(t, u(t)) \in \mathcal{M}_{\mathrm{e}}$, $t \in [t_0, t_1]$.*

Proof. The proof is similar to the proof of Theorem 3.5 and follows from Theorem 5.3 and Theorem 2.7. $\square$

Finally, using the system entropy function given by (5.26), we show a clear connection between our expanded thermodynamic model given by (5.1) and (5.2) and the entropic arrow of time.

Theorem 5.5 *Consider the large-scale dynamical system $\mathcal{G}$ with power balance equations (5.1) and (5.2) with $S(t) \equiv 0$ and $d(E, V) \equiv 0$, and assume that Axioms i) and ii) hold. Furthermore, let $s^x(\cdot, u(\cdot)) \in \mathcal{W}_{[t_0,t_1]}$, where $u(\cdot) \in \mathcal{U}$. Then for every $x_0 \notin \mathcal{M}_{\mathrm{e}}$ and $u(\cdot) \in \mathcal{U}$, there exists a continuously differentiable function $\mathcal{S} : \overline{\mathbb{R}}_+^q \times \mathbb{R}_+^q \to \mathbb{R}$ such that $\mathcal{S}(s^x(t, u(t)))$ is a strictly increasing function of time. Furthermore, $s^x(\cdot, u(\cdot))$ is an I_{2q}-reversible trajectory of $\mathcal{G}$ if and only if $s^x(t, u(t)) \in \mathcal{M}_{\mathrm{e}}$, $t \in [t_0, t_1]$.*

Proof. The existence of a continuously differentiable function $\mathcal{S} : \overline{\mathbb{R}}_+^q \times \mathbb{R}_+^q \to \mathbb{R}$, which strictly increases on all nontrivial trajectories of $\mathcal{G}$, is a restatement of Theorem 5.3 with $S(t) \equiv 0$ and $d(E, V) \equiv 0$. The proof now follows from Theorem 5.4 and Corollary 2.4. $\qquad\square$

The existence of a continuously differentiable entropy function on nontrivial irreversible trajectories of the adiabatically isolated large-scale dynamical system $\mathcal{G}$ establishes the existence of a completely ordered time set having a topological structure involving a closed set homeomorphic to the real line.

5.2 The Carnot Cycle and the Second Law of Thermodynamics

The first law of thermodynamics places no limitation on the possibility of transforming heat into work or work into heat, provided that the total amount of heat is equivalent to the total amount of work, and hence, energy is conserved in the process. The second law of thermodynamics, however, places a definite limitation on the possibility of transforming heat into work. If this were not the case, one would be able to construct a dynamical system $\mathcal{G}$, which, by extracting heat from the environment, completely transforms this heat into mechanical work. Since the supply of thermal energy contained in the universe is virtually unlimited, such a dynamical system would constitute a *perpetuum mobile* of the second kind.

There have been many statements of the second law of thermodynamics, each emphasizing another facet of the law, but all can be shown to be equivalent to one another. In this section we use the power balance equation (5.1) and (5.2) in conjunction with a *Carnotlike cycle* analysis for a large-scale dynamical system $\mathcal{G}$ to show the equivalence between the classical Kelvin-Planck and Clausius statements (postulates) of the second law of thermodynamics.

> **Kelvin-Planck.** *A transformation whose only final result is to transform completely into work heat extracted from an infinite energy source is impossible.*[1]

> **Clausius.** *A transformation whose only final result is to transfer heat from a body at a given temperature to a body at a higher temperature is impossible.*

---

[1]It is important to note that the Kelvin-Planck postulate assumes that the complete transformation of heat into work is the *only* final result of the process. It is not impossible to transform into work all the energy supplied to the system $\mathcal{G}$, provided that the state of the system is changed (see, for example, the analysis of leg $0 - 1$ in Figure 5.2).

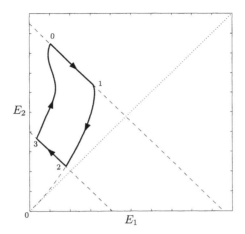

Figure 5.2 Carnot-like cycle: Leg $0-1$ (isothermal), leg $1-2$ (adiabatic), leg $2-3$ (isothermal), leg $3-0$ (adiabatic).

Note that Axiom ii) is equivalent to Clausius' postulate as applied to large-scale dynamical systems. Next, we show the equivalence of the two previous statements. First, however, we consider a Carnot-like cycle for the large-scale dynamical system $\mathcal{G}$, which consists of two *isothermal* and two *adiabatic* processes (see Figure 5.2). Recall that a process is isothermal if the total (possibly scaled) energy (temperature) of the dynamical system $\mathcal{G}$ is conserved during this process, while an adiabatic process is a process wherein no external energy in the form of heat is supplied to the system (i.e., $S(t) \equiv 0$ and $d(E, V) \equiv 0$).

First, assume the system undergoes an isothermal transformation while receiving energy (heat) from an external source and performing positive work on the environment so that the dynamics of this leg of the process are given by

$$\dot{E}(t) = w(E(t), V(t)) - d_{w0-1}(E(t), V(t)) + S_{0-1}(t), \quad E(t_0) = E_0,$$
$$t_1 \geq t \geq t_0, \quad (5.31)$$

or, since $\mathbf{e}^T w(E, V) \equiv 0$ and $\mathbf{e}^T \dot{E}(t) = 0$, $t_1 \geq t \geq t_0$,

$$0 = \mathbf{e}^T \dot{E}(t) = -\mathbf{e}^T d_{w0-1}(E(t), V(t)) + \mathbf{e}^T S_{0-1}(t), \quad t_1 \geq t \geq t_0.$$
$$(5.32)$$

Thus,

$$\mathbf{e}^T S_{0-1}(t) = \mathbf{e}^T d_{w0-1}(E(t), V(t)), \quad t_1 \geq t \geq t_0, \quad (5.33)$$

and hence, the work performed by the system during this leg of the process is, by definition,

$$L_{0-1} = \int_{t_0}^{t_1} \mathbf{e}^{\mathrm{T}} d_{\mathrm{w}0-1}(E(t), V(t)) \mathrm{dt}. \tag{5.34}$$

Moreover, it follows from (5.33) that the amount of energy (heat) supplied to the system is $Q_{0-1} = L_{0-1} > 0$.

Next, we thermally isolate the dynamical system $\mathcal{G}$ from the environment and let the system $\mathcal{G}$ perform work on the environment so that the dynamics of the adiabatic process with the initial condition $E(t_1) \in \overline{\mathbb{R}}_+^q$ are given by

$$\dot{E}(t) = w(E(t), V(t)) - d_{\mathrm{w}1-2}(E(t), V(t)), \quad t_2 \geq t \geq t_1. \tag{5.35}$$

During this leg of the process, the dynamical system $\mathcal{G}$ performs positive work on the environment by losing its internal energy so that, by (5.35),

$$\begin{aligned} L_{1-2} &= \int_{t_1}^{t_2} \mathbf{e}^{\mathrm{T}} d_{\mathrm{w}1-2}(E(t), V(t)) \mathrm{dt} \\ &= -\int_{t_1}^{t_2} \mathbf{e}^{\mathrm{T}} \dot{E}(t) \mathrm{dt} \\ &= -\Delta U \\ &> 0, \end{aligned} \tag{5.36}$$

where $\Delta U \triangleq U(t_2) - U(t_1) = \mathbf{e}^{\mathrm{T}} E(t_2) - \mathbf{e}^{\mathrm{T}} E(t_1)$ is the variation of the total energy in the system $\mathcal{G}$.

Next, we perform an isothermal process during which positive work is being performed on the system $\mathcal{G}$ while the system surrenders energy to the environment. The dynamics of this leg of the process are characterized by

$$\dot{E}(t) = w(E(t), V(t)) + S_{\mathrm{w}2-3}(t) - d_{2-3}(E(t), V(t)), \quad E(t_2) \in \overline{\mathbb{R}}_+^q,$$
$$t_3 \geq t \geq t_2, \tag{5.37}$$

where $\mathbf{e}^{\mathrm{T}} \dot{E}(t) \equiv 0$. Thus, the work done by the system $\mathcal{G}$ during this transformation is negative and is given by

$$\begin{aligned} L_{2-3} &= -\int_{t_2}^{t_3} \mathbf{e}^{\mathrm{T}} S_{\mathrm{w}2-3}(t) \mathrm{dt} \\ &= -\int_{t_2}^{t_3} \mathbf{e}^{\mathrm{T}} d_{2-3}(E(t), V(t)) \mathrm{dt} \\ &= -Q_{2-3} \\ &< 0, \end{aligned} \tag{5.38}$$

where $Q_{2-3} > 0$ is the amount of energy released to the environment.

Finally, we perform an adiabatic process to drive the system back to its initial state $E_0 \in \overline{\mathbb{R}}_+^q$, $V_0 \in \mathbb{R}_+^q$. During this process, work is performed on the system by the environment so that the dynamics of this leg of the process are given by

$$\dot{E}(t) = w(E(t), V(t)) + S_{w3-0}(t), \quad E(t_3) \in \overline{\mathbb{R}}_+^q, \quad t_4 \geq t \geq t_3. \quad (5.39)$$

The work done by the dynamical system $\mathcal{G}$ during this transformation is negative and is given by

$$L_{3-0} = -\int_{t_3}^{t_4} \mathbf{e}^{\mathrm{T}} S_{w3-0}(t) \mathrm{d}t = -\int_{t_3}^{t_4} \mathbf{e}^{\mathrm{T}} \dot{E}(t) \mathrm{d}t = \Delta U < 0. \quad (5.40)$$

Thus, it follows from (5.5) that the total work done by the system during the cycle is given by $L = Q_{0-1} - Q_{2-3}$. This implies that only part of the heat Q_{0-1} that is absorbed by the large-scale dynamical system from the external source at a higher energy level is transformed into work by the Carnot-like cycle; the rest of the heat Q_{2-3} is surrendered to the external source at a lower energy level. Assuming an infinite energy source and repeating this cycle arbitrarily often establishes the impossibility of a *perpetuum mobile* of the second kind. Finally, since the *efficiency* η of a Carnot-like cycle is given by the ratio of the work done by the system and the energy supplied throughout the cycle, it follows that

$$\eta = \frac{Q_{0-1} - Q_{2-3}}{Q_{0-1}} = 1 - \frac{Q_{2-3}}{Q_{0-1}}. \quad (5.41)$$

To show the equivalence between the Kelvin-Planck and the Clausius statements of the second law of thermodynamics, we use a *reductio ad absurdum* argument along with a contrapositive argument, that is, if one of the statements were not valid, then the other statement would also not be valid, and vice versa. Hence, suppose, *ad absurdum*, that in an isolated large-scale system energy can flow from less energetic subsystems to more energetic subsystems. Then, for a Carnot-like cycle we would be able to perform the last isothermal (leg $2 - 3$ in Figure 5.2) transformation without doing any external work on the system $\mathcal{G}$. Specifically, in this case the dynamics of the isothermal process would be given by

$$\dot{E}(t) = w(E(t), V(t)) - d_{2-3}(E(t), V(t)), \quad t_3 \geq t \geq t_2, \quad (5.42)$$

and hence,

$$0 = \mathbf{e}^{\mathrm{T}} \dot{E}(t) = -\mathbf{e}^{\mathrm{T}} d_{2-3}(E(t), V(t)), \quad t_3 \geq t \geq t_2, \quad (5.43)$$

which implies that

$$Q_{2-3} = \int_{t_2}^{t_3} \mathbf{e}^{\mathrm{T}} d_{2-3}(E(t), V(t)) \mathrm{d}t = 0. \tag{5.44}$$

In this case, the efficiency of the Carnot cycle, given by (5.41), is $\eta = 1$. Hence, it would be possible to transform into work all the energy (heat) absorbed by the large-scale system $\mathcal{G}$ without producing any other change in the system. This contradicts the Kelvin-Planck postulate.

Conversely, suppose, *ad absurdum*, that it were possible to transform completely into work an amount of energy (heat) supplied to the system $\mathcal{G}$ by an infinite heat source with no other changes in the system. Then it follows from (5.41) that $Q_{2-3} = 0$, and since $d_{2-3}(E(t), V(t)) \geq\geq 0$, $t_3 \geq t \geq t_2$, it follows that $d_{2-3}(E(t), V(t)) \equiv 0$. Moreover, since $S_{\mathrm{w}2-3}(t) \geq\geq 0$, $t_3 \geq t \geq t_2$, it follows from (5.38) that $S_{\mathrm{w}2-3}(t) \equiv 0$. Here, it suffices to consider the case of two subsystems to arrive at a contradiction to Clausius' postulate. In this particular case, the dynamics characterized by (5.37) of the second isothermal process become

$$\dot{E}(t) = w(E(t), V(t)), \quad t_3 \geq t \geq t_2. \tag{5.45}$$

Premultiplying (5.45) by $E^{\mathrm{T}}(t)$, $t_3 \geq t \geq t_2$, and using the fact that during this isothermal transformation $\dot{E}_1(t) = -\dot{E}_2(t) \leq 0$, $t_3 \geq t \geq t_2$, and $E_1(t) \leq E_2(t)$, $t_3 \geq t \geq t_2$, (see Figure 5.2), it follows that for $t_3 \geq t \geq t_2$,

$$[E_1(t) - E_2(t)]\phi_{12}(E(t), V(t)) = [E_1(t) - E_2(t)]\dot{E}_1(t) \geq 0, \tag{5.46}$$

which implies that during this isothermal transformation energy flows from the less energetic subsystem $\mathcal{G}_1$ (cooler object) to the more energetic subsystem $\mathcal{G}_2$ (hotter object). This contradicts Clausius' postulate. Thus, the Kelvin-Planck and Clausius statements of the second law of thermodynamics are equivalent.

Chapter Six

Thermodynamic Systems with Linear Energy Exchange

6.1 Linear Thermodynamic System Models

In this chapter we specialize the results of Chapter 3 to the case of large-scale dynamical systems with linear energy exchange between subsystems, that is, $w(E) = WE$ and $d(E) = DE$, where $W \in \mathbb{R}^{q \times q}$ and $D \in \mathbb{R}^{q \times q}$. In this case, the vector form of the energy balance equation (3.1), with $t_0 = 0$, is given by

$$E(T) = E(0) + \int_0^T WE(t)\mathrm{d}t - \int_0^T DE(t)\mathrm{d}t + \int_0^T S(t)\mathrm{d}t, \quad T \geq 0,$$

$$(6.1)$$

or, in power balance form,

$$\dot{E}(t) = WE(t) - DE(t) + S(t), \quad E(0) = E_0, \quad t \geq 0. \quad (6.2)$$

Next, let the net energy flow from the jth subsystem $\mathcal{G}_j$ to the ith subsystem $\mathcal{G}_i$ be parameterized as $\phi_{ij}(E) = \Phi_{ij}^{\mathrm{T}}E$, where $\Phi_{ij} \in \mathbb{R}^q$ and $E \in \overline{\mathbb{R}}_+^q$. In this case, since $w_i(E) = \sum_{i=1, j \neq i}^q \phi_{ij}(E)$, it follows that

$$W = \left[\sum_{j=2}^q \Phi_{1j}, \ldots, \sum_{j=1, j \neq i}^q \Phi_{ij}, \ldots, \sum_{j=1}^{q-1} \Phi_{qj} \right]^{\mathrm{T}}. \quad (6.3)$$

Since $\phi_{ij}(E) = -\phi_{ji}(E)$, $i, j = 1, ..., q$, $i \neq j$, and $E \in \overline{\mathbb{R}}_+^q$, it follows that $\Phi_{ij} = -\Phi_{ji}$, $i \neq j$, $i, j = 1, ..., q$. The following proposition gives necessary and sufficient conditions on W so that Axioms $i)$ and $ii)$ hold.

Proposition 6.1 *Consider the large-scale dynamical system $\mathcal{G}$ with power balance equation given by (6.2) and with $D = 0$. Then Axioms $i)$ and $ii)$ hold if and only if $W = W^{\mathrm{T}}$, $W\mathbf{e} = 0$, $\mathrm{rank}\, W = q - 1$, and W is essentially nonnegative.*

Proof. Assume Axioms i) and ii) hold. Since by Axiom ii) $(E_i - E_j)\phi_{ij}(E) \leq 0$, $E \in \overline{\mathbb{R}}_+^q$, it follows that $E^T\Phi_{ij}\mathbf{e}_{ij}^T E \leq 0$, $i,j = 1,...,q$, $i \neq j$, where $E \in \overline{\mathbb{R}}_+^q$ and $\mathbf{e}_{ij} \in \mathbb{R}^q$ is a vector whose ith entry is 1, jth entry is -1, and remaining entries are zero. Next, it can be shown that $E^T\Phi_{ij}\mathbf{e}_{ij}^T E \leq 0$, $E \in \overline{\mathbb{R}}_+^q$, $i \neq j$, $i,j = 1,...,q$, if and only if $\Phi_{ij} \in \mathbb{R}^q$ is such that its ith entry is $-\sigma_{ij}$, its jth entry is σ_{ij}, where $\sigma_{ij} \geq 0$, and its remaining entries are zero. Furthermore, since $\Phi_{ij} = -\Phi_{ji}$, $i \neq j$, $i,j = 1,...,q$, it follows that $\sigma_{ij} = \sigma_{ji}$, $i \neq j$, $i,j = 1,...,q$. Hence, W is given by

$$W_{(i,j)} = \begin{cases} -\sum_{k=1,\,k\neq j}^q \sigma_{kj}, & i = j, \\ \sigma_{ij}, & i \neq j, \end{cases} \tag{6.4}$$

which implies that W is symmetric (since $\sigma_{ij} = \sigma_{ji}$), essentially non-negative, and $W\mathbf{e} = 0$. Now, since by Axiom i) $\phi_{ij}(E) = 0$ if and only if $E_i = E_j$ for all $i,j = 1,...,q$, $i \neq j$, such that $\mathcal{C}_{(i,j)} = 1$, it follows that $\sigma_{ij} > 0$ for all $i,j = 1,...,q$, $i \neq j$, such that $\mathcal{C}_{(i,j)} = 1$. Hence, rank $W =$ rank $\mathcal{C} = q - 1$.

The converse is immediate and hence is omitted. $\square$

Next, we specialize the energy balance equation (6.2) to the case where $D = \mathrm{diag}[\sigma_{11}, \sigma_{22}, ..., \sigma_{qq}]$. In this case, the vector form of the energy balance equation (6.1) is given by

$$E(T) = E(0) + \int_0^T AE(t)\mathrm{d}t + \int_0^T S(t)\mathrm{d}t, \quad T \geq 0, \tag{6.5}$$

or, in power balance form,

$$\dot{E}(t) = AE(t) + S(t), \quad E(0) = E_0, \quad t \geq 0, \tag{6.6}$$

where $A \triangleq W - D$ is such that

$$A_{(i,j)} = \begin{cases} -\sum_{k=1}^q \sigma_{kj}, & i = j, \\ \sigma_{ij}, & i \neq j. \end{cases} \tag{6.7}$$

Note that (6.7) implies $\sum_{i=1}^q A_{(i,j)} = -\sigma_{jj} \leq 0$, $j = 1,...,q$, and hence, A is a semistable compartmental matrix (see iv) of Lemma 6.1). If $\sigma_{ii} > 0$, $i = 1,...,q$, then A is an asymptotically stable compartmental matrix.

An important special case of (6.6) is the case where A is symmetric or, equivalently, $\sigma_{ij} = \sigma_{ji}$, $i \neq j$, $i,j = 1,...,q$. In this case, it follows from (6.6) that for each subsystem the power balance equation satisfies

$$\dot{E}_i(t) + \sigma_{ii}E_i(t) + \sum_{j=1,\,j\neq i}^q \sigma_{ij}[E_i(t) - E_j(t)] = S_i(t), \quad t \geq 0. \tag{6.8}$$

Note that $\phi_i(E) \triangleq \sum_{j=1, j\neq i}^q \sigma_{ij}[E_i - E_j]$, $E \in \overline{\mathbb{R}}_+^q$, $i = 1, ..., q$, represents the energy flow from the ith subsystem to all other subsystems and is given by the sum of the individual energy flows from the ith subsystem to the jth subsystem. Furthermore, these energy flows are proportional to the energy differences of the subsystems, that is, $E_i - E_j$. Hence, (6.8) is a power balance equation that governs the energy exchange among coupled subsystems and is completely analogous to the equations of thermal transfer with subsystem energies playing the role of temperatures. Furthermore, note that since $\sigma_{ij} \geq 0$, $i \neq j$, $i, j = 1, ..., q$, energy flows from more energetic subsystems to less energetic subsystems, which is consistent with the second law of thermodynamics requiring that heat (energy) *must* flow in the direction of lower temperatures.

The next lemma and proposition are needed for developing expressions for steady-state energy distributions of the large-scale dynamical system $\mathcal{G}$ with the linear power balance equation (6.6).

Lemma 6.1 *Let $A \in \mathbb{R}^{q \times q}$ be compartmental and $S \in \mathbb{R}^q$. Then the following properties hold:*

i) $-A$ *is an M-matrix.*

ii) If $\lambda \in \mathrm{spec}(A)$*, then either* $\mathrm{Re}\,\lambda < 0$ *or* $\lambda = 0$*.*

iii) $\mathrm{ind}\,(A) \leq 1$*.*

iv) A is semistable and $\lim_{t\to\infty} e^{At} = I_q - AA^{\#} \geq\geq 0$*.*

v) $\mathcal{R}(A) = \mathcal{N}(I_q - AA^{\#})$ *and* $\mathcal{N}(A) = \mathcal{R}(I_q - AA^{\#})$*.*

vi) $\int_0^t e^{As}\mathrm{d}s = A^{\#}(e^{At} - I_q) + (I_q - AA^{\#})t$*,* $t \geq 0$*.*

vii) $\int_0^\infty e^{At}\mathrm{d}t\, S$ *exists if and only if* $S \in \mathcal{R}(A)$*, where* $S \in \mathbb{R}^q$*.*

viii) If $S \in \mathcal{R}(A)$*, then* $\int_0^\infty e^{At}\mathrm{d}t\, S = -A^{\#}S$*.*

ix) If $S \in \mathcal{R}(A)$ *and* $S \geq\geq 0$*, then* $-A^{\#}S \geq\geq 0$*.*

x) A is nonsingular if and only if $-A$ is a nonsingular M-matrix.

xi) If A is nonsingular, then A is asymptotically stable and $-A^{-1} \geq\geq 0$*.*

Proof. The proof of the result appears in [10]. For completeness of exposition, we provide a proof here.

i) Since, by (6.7), $-A^{\mathrm{T}}\mathbf{e} \geq\geq 0$, and $-A$ is a Z-matrix, it follows from Theorem 1 of [7] that $-A^{\mathrm{T}}$, and hence, $-A$ is an M-matrix.

ii) Since $-A$ is an M-matrix, it follows from Theorem 4.6 of [6, p. 150], that Re $\lambda < 0$ or $\lambda = 0$, where $\lambda \in \mathrm{spec}(A)$.

iii) This follows from the fact that since $-A^{\mathrm{T}}\mathbf{e} \geq\geq 0$, then $-A$ has "property c" (see [6, p. 152 and 155]). Hence, since an M-matrix $-A \in \mathbb{R}^{q \times q}$ has "property c" if and only if $\mathrm{ind}(-A) \leq 1$ (see [6, Lemma 4.11, p. 153]), it follows that $\mathrm{ind}(-A) = \mathrm{ind}(A) \leq 1$.

iv) Since $\mathrm{ind}(A) \leq 1$, it follows from the real Jordan decomposition that there exist invertible matrices $J \in \mathbb{R}^{r \times r}$, where $r = \mathrm{rank}\ A$, and $T \in \mathbb{R}^{q \times q}$ such that

$$A = T \begin{bmatrix} J & 0 \\ 0 & 0 \end{bmatrix} T^{-1} \tag{6.9}$$

and J is asymptotically stable. Hence, it follows that

$$\begin{aligned}
\lim_{t \to \infty} e^{At} &= \lim_{t \to \infty} T \begin{bmatrix} e^{Jt} & 0 \\ 0 & I_{q-r} \end{bmatrix} T^{-1} \\
&= T \begin{bmatrix} 0 & 0 \\ 0 & I_{q-r} \end{bmatrix} T^{-1} \\
&= I_q - T \begin{bmatrix} J & 0 \\ 0 & 0 \end{bmatrix} T^{-1} T \begin{bmatrix} J^{-1} & 0 \\ 0 & 0 \end{bmatrix} T^{-1} \\
&= I_q - AA^{\#}. \tag{6.10}
\end{aligned}$$

Next, since A is essentially nonnegative, it follows from Corollary 2.1 that $e^{At} \geq\geq 0$, $t \geq 0$, which implies that $I_q - AA^{\#} \geq\geq 0$.

v) Let $x \in \mathcal{R}(A)$, that is, there exists $y \in \mathbb{R}^q$ such that $x = Ay$. Now, $(I - AA^{\#})x = x - AA^{\#}Ay = x - Ay = 0$, which implies that $\mathcal{R}(A) \subseteq \mathcal{N}(I - AA^{\#})$. Conversely, let $x \in \mathcal{N}(I - AA^{\#})$. Hence, $(I - AA^{\#})x = 0$ or, equivalently, $x = AA^{\#}x$, which implies that $x \in \mathcal{R}(A)$, and hence, proves $\mathcal{R}(A) = \mathcal{N}(I - AA^{\#})$. The equality $\mathcal{N}(A) = \mathcal{R}(I - AA^{\#})$ can be proved in an analogous manner.

vi) Note that $A = T \begin{bmatrix} J & 0 \\ 0 & 0 \end{bmatrix} T^{-1}$, and hence,

$$\begin{aligned}
\int_0^t e^{A\sigma} \mathrm{d}\sigma &= \int_0^t T \begin{bmatrix} e^{J\sigma} & 0 \\ 0 & I_{q-r} \end{bmatrix} T^{-1} \mathrm{d}\sigma \\
&= T \begin{bmatrix} \int_0^t e^{J\sigma} \mathrm{d}\sigma & 0 \\ 0 & \int_0^t I_{q-r} \mathrm{d}\sigma \end{bmatrix} T^{-1} \\
&= T \begin{bmatrix} J^{-1}(e^{Jt} - I_r) & 0 \\ 0 & I_{q-r}t \end{bmatrix} T^{-1}
\end{aligned}$$

$$= T \begin{bmatrix} J^{-1} & 0 \\ 0 & 0 \end{bmatrix} T^{-1} T \begin{bmatrix} (e^{Jt} - I_r) & 0 \\ 0 & -I_{q-r} \end{bmatrix} T^{-1}$$

$$+ T \begin{bmatrix} 0 & 0 \\ 0 & I_{q-r}t \end{bmatrix} T^{-1}$$

$$= A^{\#}(e^{At} - I_q) + (I_q - AA^{\#})t, \quad t \geq 0. \tag{6.11}$$

vii) The result is a direct consequence of iv), v), and vi).

$viii$) The result is a direct consequence of iv) and vi).

ix) The result follows from $viii$) and the fact that $e^{At} \geq\geq 0$, $t \geq 0$.

x) The result follows from i).

xi) Asymptotic stability of A is a direct consequence of ii), while $A^{-1} \leq\leq 0$ follows from ix) with $S = \text{col}_i(I_q)$, $i = 1, ..., q$, where $\text{col}_i(I_q)$ denotes the ith column of I_q. $\square$

Proposition 6.2 *Consider the large-scale dynamical system $\mathcal{G}$ with power balance equation given by (6.6). Suppose $E_0 \geq\geq 0$ and $S(t) \geq\geq 0$, $t \geq 0$. Then the solution $E(t)$ to (6.6) is nonnegative for all $t \geq 0$ if and only if A is essentially nonnegative.*

Proof. It follows from Lagrange's formula that the solution $E(t)$, $t \geq 0$, to (6.6) is given by

$$E(t) = e^{At}E(0) + \int_0^t e^{A(t-\sigma)} S(\sigma) d\sigma, \quad t \geq 0. \tag{6.12}$$

Now, if A is essentially nonnegative, it follows from Corollary 2.1 that $e^{At} \geq\geq 0$, $t \geq 0$, and if $E(0) \in \mathbb{R}_+^q$ and $S(t) \geq\geq 0$, $t \geq 0$, then it follows that $E(t) \geq\geq 0$ for all $t \geq 0$.

Conversely, suppose that the solution $E(t)$, $t \geq 0$, to (6.6) is nonnegative for all $E_0 \geq\geq 0$. Then, with $S(t) \equiv 0$, $E(t) = e^{At}E_0$, and hence, e^{At} is nonnegative for all $t \geq 0$. Thus, it follows from Corollary 2.1 that A is essentially nonnegative, which proves the result. $\square$

Next, we develop expressions for the steady-state energy distribution for the large-scale dynamical system $\mathcal{G}$ for the cases where supplied system power $S(t)$ is a periodic function with period $\tau > 0$ (that is, $S(t + \tau) = S(t)$, $t \geq 0$) and $S(t)$ is constant (that is, $S(t) \equiv S$). Define $e(t) \triangleq E(t) - E(t + \tau)$, $t \geq 0$, and note that

$$\dot{e}(t) = Ae(t), \quad e(0) = E(0) - E(\tau), \quad t \geq 0. \tag{6.13}$$

Hence, since

$$e(t) = e^{At}[E(0) - E(\tau)], \quad t \geq 0, \tag{6.14}$$

and A is semistable, it follows from $iv)$ of Lemma 6.1 that

$$\lim_{t\to\infty} e(t) = \lim_{t\to\infty} [E(t) - E(t+\tau)] = (I_q - AA^\#)[E(0) - E(\tau)], \quad (6.15)$$

which represents a constant offset to the steady-state error energy distribution in the large-scale dynamical system $\mathcal{G}$. For the case where $S(t) \equiv S$, $\tau \to \infty$, and hence, the following result is immediate. This result first appeared in [10].

Proposition 6.3 *Consider the large-scale dynamical system $\mathcal{G}$ with power balance equation given by (6.6). Suppose that $E_0 \geq\geq 0$ and $S(t) \equiv S \geq\geq 0$. Then $E_\infty \triangleq \lim_{t\to\infty} E(t)$ exists if and only if $S \in \mathcal{R}(A)$. In this case,*

$$E_\infty = (I_q - AA^\#)E_0 - A^\#S \qquad (6.16)$$

and $E_\infty \geq\geq 0$. If, in addition, A is nonsingular, then E_∞ exists for all $S \geq\geq 0$ and is given by

$$E_\infty = -A^{-1}S. \qquad (6.17)$$

   Proof. Note that it follows from Lagrange's formula that the solution $E(t)$, $t \geq 0$, to (6.6) is given by

$$E(t) = e^{At}E(0) + \int_0^t e^{A(t-\sigma)}Sd\sigma, \quad t \geq 0. \qquad (6.18)$$

Now, the result is a direct consequence of Proposition 6.2 and $iv)$, $vii)$, $viii)$, and $ix)$ of Lemma 6.1. $\square$

6.2 Semistability and Energy Equipartition in Linear Thermodynamic Models

In this section, we show that an isolated large-scale linear dynamical system as well as a nonisolated large-scale linear dynamical system with strong coupling between subsystems and a constant heat flux input has a tendency to uniformly distribute its energy among all of its parts. First, we begin by specializing the result of Proposition 6.3 to the case where there is no energy dissipation from each subsystem $\mathcal{G}_i$ of $\mathcal{G}$, that is, $\sigma_{ii} = 0$, $i = 1, ..., q$. Note that in this case $\mathbf{e}^T A = 0$, and hence, $\operatorname{rank} A \leq q - 1$. Furthermore, if $S = 0$, it follows from (6.6) that $\mathbf{e}^T \dot{E}(t) = \mathbf{e}^T AE(t) = 0$, $t \geq 0$, and hence, the total energy of the isolated large-scale dynamical system $\mathcal{G}$ is conserved.

Theorem 6.1 *Consider the large-scale dynamical system $\mathcal{G}$ with power balance equation given by (6.6). Assume* $\operatorname{rank} A = q - 1$, $\sigma_{ii} = 0$, $i = 1, ..., q$, *and* $A = A^{\mathrm{T}}$. *If* $E_0 \geq\geq 0$ *and* $S(t) \equiv 0$, *then the equilibrium state* $\alpha \mathbf{e}$, $\alpha \geq 0$, *of the isolated system* $\mathcal{G}$ *is semistable and the steady-state energy distribution* E_∞ *of the isolated large-scale dynamical system* $\mathcal{G}$ *is given by*

$$E_\infty = \left[\frac{1}{q} \sum_{i=1}^{q} E_{i0} \right] \mathbf{e}. \tag{6.19}$$

If, in addition, for some $k \in \{1, ..., q\}$, $\sigma_{kk} > 0$, *then the zero solution* $E(t) \equiv 0$ *to (6.6) is globally asymptotically stable.*

Proof. Note that since $\mathbf{e}^{\mathrm{T}} A = 0$ it follows from (6.6) with $S(t) \equiv 0$ that $\mathbf{e}^{\mathrm{T}} \dot{E}(t) = 0$, $t \geq 0$, and hence, $\mathbf{e}^{\mathrm{T}} E(t) = \mathbf{e}^{\mathrm{T}} E_0$, $t \geq 0$. Furthermore, since by Proposition 6.2 the solution $E(t)$, $t \geq t_0$, to (6.6) is nonnegative, it follows that $0 \leq E_i(t) \leq \mathbf{e}^{\mathrm{T}} E(t) = \mathbf{e}^{\mathrm{T}} E_0$, $t \geq 0$, $i = 1, ..., q$. Hence, the solution $E(t)$, $t \geq 0$, to (6.6) is bounded for all $E_0 \in \overline{\mathbb{R}}_+^q$. Next, note that $\phi_{ij}(E) = \sigma_{ij}(E_j - E_i)$ and $(E_i - E_j)\phi_{ij}(E) = -\sigma_{ij}(E_i - E_j)^2 \leq 0$, $E \in \overline{\mathbb{R}}_+^q$, $i \neq j$, $i, j = 1, ..., q$, which implies that Axioms i) and ii) are satisfied, and hence, $\mathcal{G}$ is a thermodynamically consistent linear energy flow model. Thus, $E = \alpha \mathbf{e}$, $\alpha \geq 0$, is the equilibrium state of the isolated large-scale dynamical system $\mathcal{G}$. To show Lyapunov stability of the equilibrium state $\alpha \mathbf{e}$, consider the shifted-system ectropy function $\mathcal{E}_{\mathrm{s}}(E) = \frac{1}{2}(E - \alpha \mathbf{e})^{\mathrm{T}}(E - \alpha \mathbf{e})$, $E \in \overline{\mathbb{R}}_+^q$, as a Lyapunov function candidate. Then the Lyapunov derivative is given by

$$\begin{aligned}
\dot{\mathcal{E}}_{\mathrm{s}}(E) &= (E - \alpha \mathbf{e})^{\mathrm{T}} A E \\
&= E^{\mathrm{T}} A E \\
&= -\sum_{i=1}^{q} \sum_{j=i+1}^{q} \sigma_{ij}(E_i - E_j)^2 \\
&\leq 0, \quad E \in \overline{\mathbb{R}}_+^q, \tag{6.20}
\end{aligned}$$

which implies Lyapunov stability of the equilibrium state $\alpha \mathbf{e}$, $\alpha \geq 0$.

Next, consider the set $\mathcal{R} \triangleq \{E \in \overline{\mathbb{R}}_+^q : \dot{\mathcal{E}}_{\mathrm{s}}(E) = 0\} = \{E \in \overline{\mathbb{R}}_+^q : E^{\mathrm{T}} A E = 0\}$. Since A is compartmental and symmetric, it follows from ii) of Lemma 6.1 that A is a negative semi-definite matrix, and hence, $E^{\mathrm{T}} A E = 0$ if and only if $A E = 0$. Since, by assumption $\operatorname{rank} A = q - 1$, it follows that there exists one and only one linearly independent solution to $A E = 0$ given by $E = \mathbf{e}$. Hence, $\mathcal{R} = \{E \in \overline{\mathbb{R}}_+^q : E = \alpha \mathbf{e}, \alpha \geq 0\}$. Since $\mathcal{R}$ consists of only equilibrium states of (6.6), it follows that $\mathcal{M} = \mathcal{R}$, where $\mathcal{M}$ is the largest invariant set contained in $\mathcal{R}$. Hence, for every $E_0 \in \overline{\mathbb{R}}_+^q$, it follows from the Krasovskii-LaSalle invariant set theorem that $E(t) \to \alpha \mathbf{e}$ as $t \to \infty$ for some $\alpha \geq 0$, and hence, $\alpha \mathbf{e}$, $\alpha \geq 0$, is a semistable equilibrium state of (6.6). Furthermore, since the energy is conserved in the isolated large-scale dynamical system $\mathcal{G}$, it follows that $q\alpha = \mathbf{e}^\mathrm{T} E_0$. Thus, $\alpha = \frac{1}{q} \sum_{i=1}^{q} E_{i0}$, which implies (6.19).

Finally, to show that in case where $\sigma_{kk} > 0$ for some $k \in \{1, ..., q\}$, the zero solution $E(t) \equiv 0$ to (6.6) is globally asymptotically stable, consider the system ectropy $\mathcal{E}(E) = \frac{1}{2} E^\mathrm{T} E$, $E \in \overline{\mathbb{R}}_+^q$, as a Lyapunov function candidate. Note that Lyapunov stability of the zero equilibrium state follows from the previous analysis with $\alpha = 0$. Next, the Lyapunov derivative is given by

$$\dot{\mathcal{E}}(E) = E^\mathrm{T} A E = -\sum_{i=1}^{q} \sum_{j=i+1}^{q} \sigma_{ij}(E_i - E_j)^2 - \sigma_{kk} E_k^2, \quad E \in \overline{\mathbb{R}}_+^q. \tag{6.21}$$

Consider the set $\mathcal{R} \triangleq \{E \in \overline{\mathbb{R}}_+^q : \dot{\mathcal{E}}(E) = 0\} = \{E \in \overline{\mathbb{R}}_+^q : E_1 = \cdots = E_q\} \cap \{E \in \overline{\mathbb{R}}_+^q : E_k = 0, k \in \{1, ..., q\}\} = \{0\}$. Hence, the largest invariant set contained in $\mathcal{R}$ is given by $\mathcal{M} = \mathcal{R} = \{0\}$. Thus, it follows from the Krasovskii-LaSalle invariant set theorem that $E(t) \to \mathcal{M} = \{0\}$ as $t \to \infty$, which proves that the zero solution $E(t) \equiv 0$ to (6.6) is globally asymptotically stable. $\qquad\square$

The result of Theorem 6.1 can also be obtained as a direct consequence of Theorem 3.9 with $w(E) = WE$ and $d(E) = DE$. To see this, note that it follows from Proposition 6.1 that the symmetry condition $W = W^\mathrm{T}$ along with $W\mathbf{e} = 0$ and $\operatorname{rank} W = q - 1$ ensure that Axioms $i)$ and $ii)$ are satisfied for the linear energy flow model (6.6). Furthermore, the condition $\operatorname{rank} W = q - 1$ ensures that the directed graph associated with the connectivity matrix $\mathcal{C}$ for $\mathcal{G}$ is strongly connected. The result now follows from Theorem 3.9.

Finally, we examine the steady-state energy distribution for large-scale linear dynamical systems $\mathcal{G}$ in case of strong coupling between subsystems, that is, $\sigma_{ij} \to \infty$, $i \neq j$, $i, j = 1, ..., q$. For this analysis we assume that A given by (6.7) is symmetric, that is, $\sigma_{ij} = \sigma_{ji}$, $i \neq j$, $i, j = 1, ..., q$, and $\sigma_{ii} > 0$, $i = 1, ..., q$. Thus, $-A$ is a nonsingular M-matrix for all values of σ_{ij}, $i \neq j$, $i, j = 1, ..., q$. Moreover, in this case it follows that if $\frac{\sigma_{ij}}{\sigma_{kl}} \to 1$ as $\sigma_{ij} \to \infty$, $i \neq j$, and $\sigma_{kl} \to \infty$, $k \neq l$, then

$$\lim_{\sigma_{ij} \to \infty, \, i \neq j} A^{-1} = \lim_{\sigma \to \infty} [-D + \sigma(-q I_q + \mathbf{e}\mathbf{e}^\mathrm{T})]^{-1}, \tag{6.22}$$

where $D = \mathrm{diag}[\sigma_{11}, ..., \sigma_{qq}] > 0$. The following lemmas are needed for the next result.

Lemma 6.2 *Let* $Y \in \mathbb{R}^{q \times q}$ *be such that* $\mathrm{ind}\,(Y) \le 1$. *Then* $\lim_{\sigma \to \infty} (I_q - \sigma Y)^{-1} = I_q - Y^{\#} Y$.

Proof. Note that

$$(I_q - \sigma Y)^{-1} = I_q + \sigma (I_q - \sigma Y)^{-1} Y$$

$$= I_q + \left(\frac{1}{\sigma} I_q - Y \right)^{-1} Y$$

$$= I_q - \left(Y - \frac{1}{\sigma} I_q \right)^{-1} Y. \tag{6.23}$$

Now, using the fact that if $N \in \mathbb{R}^{q \times q}$ and $\mathrm{ind}\, N \le 1$, then

$$\lim_{\alpha \to 0} (N + \alpha I)^{-1} N = N N^{\#} = N^{\#} N, \tag{6.24}$$

it follows that

$$\lim_{\sigma \to \infty} (I_q - \sigma Y)^{-1} = I_q - \lim_{\frac{1}{\sigma} \to 0} \left(Y - \frac{1}{\sigma} I_q \right)^{-1} Y = I_q - Y^{\#} Y, \tag{6.25}$$

which proves the result. $\qquad\square$

Lemma 6.3 *Let* $D \in \mathbb{R}^{q \times q}$ *and* $X \in \mathbb{R}^{q \times q}$ *be such that* $D > 0$ *and* $X = -q I_q + \mathbf{e}\mathbf{e}^{\mathrm{T}}$. *Then*

$$I_q - Y^{\#} Y = \frac{D^{\frac{1}{2}} \mathbf{e}\mathbf{e}^{\mathrm{T}} D^{\frac{1}{2}}}{\mathbf{e}^{\mathrm{T}} D \mathbf{e}}, \tag{6.26}$$

where $Y \triangleq D^{-\frac{1}{2}} X D^{-\frac{1}{2}}$.

Proof. Note that

$$Y = D^{-\frac{1}{2}} (-q I_q + \mathbf{e}\mathbf{e}^{\mathrm{T}}) D^{-\frac{1}{2}} = -q D^{-1} + D^{-\frac{1}{2}} \mathbf{e}\mathbf{e}^{\mathrm{T}} D^{-\frac{1}{2}}. \tag{6.27}$$

Now, using the fact that if $N \in \mathbb{R}^{q \times q}$ is nonsingular and symmetric and $b \in \mathbb{R}^q$ is a nonzero vector, then [74]

$$(N + bb^{\mathrm{T}})^{\#}$$

$$= \left(I - \frac{1}{b^{\mathrm{T}} N^{-2} b} N^{-1} bb^{\mathrm{T}} N^{-1} \right) N^{-1} \left(I - \frac{1}{b^{\mathrm{T}} N^{-2} b} N^{-1} bb^{\mathrm{T}} N^{-1} \right),$$

$$\tag{6.28}$$

it follows that

$$-Y^{\#} = \frac{1}{q}\left(I_q - \frac{D^{\frac{1}{2}}\mathbf{e}\mathbf{e}^{\mathrm{T}}D^{\frac{1}{2}}}{\mathbf{e}^{\mathrm{T}}D\mathbf{e}}\right)D\left(I_q - \frac{D^{\frac{1}{2}}\mathbf{e}\mathbf{e}^{\mathrm{T}}D^{\frac{1}{2}}}{\mathbf{e}^{\mathrm{T}}D\mathbf{e}}\right). \quad (6.29)$$

Hence,

$$\begin{aligned}
-Y^{\#}Y &= -\left(I_q - \frac{D^{\frac{1}{2}}\mathbf{e}\mathbf{e}^{\mathrm{T}}D^{\frac{1}{2}}}{\mathbf{e}^{\mathrm{T}}D\mathbf{e}}\right)D\left(I_q - \frac{D^{\frac{1}{2}}\mathbf{e}\mathbf{e}^{\mathrm{T}}D^{\frac{1}{2}}}{\mathbf{e}^{\mathrm{T}}D\mathbf{e}}\right) \\
&\quad \cdot \left(D^{-1} - \frac{1}{q}D^{-\frac{1}{2}}\mathbf{e}\mathbf{e}^{\mathrm{T}}D^{-\frac{1}{2}}\right) \\
&= -\left(I_q - \frac{D^{\frac{1}{2}}\mathbf{e}\mathbf{e}^{\mathrm{T}}D^{\frac{1}{2}}}{\mathbf{e}^{\mathrm{T}}D\mathbf{e}}\right). \quad (6.30)
\end{aligned}$$

Thus, $I_q - Y^{\#}Y = \frac{D^{\frac{1}{2}}\mathbf{e}\mathbf{e}^{\mathrm{T}}D^{\frac{1}{2}}}{\mathbf{e}^{\mathrm{T}}D\mathbf{e}}$. $\square$

Proposition 6.4 *Consider the large-scale dynamical system $\mathcal{G}$ with power balance equation given by (6.6). Let $S(t) \equiv S$, $S \in \mathbb{R}^{q \times q}$, let $A \in \mathbb{R}^{q \times q}$ be compartmental, and assume A is symmetric, $\sigma_{ii} > 0$, $i = 1, ..., q$, and $\frac{\sigma_{ij}}{\sigma_{kl}} \to 1$ as $\sigma_{ij} \to \infty$, $i \neq j$, and $\sigma_{kl} \to \infty$, $k \neq l$. Then the steady-state energy distribution E_∞ of the large-scale dynamical system $\mathcal{G}$ is given by*

$$E_\infty = \left[\frac{\mathbf{e}^{\mathrm{T}}S}{\sum_{i=1}^{q}\sigma_{ii}}\right]\mathbf{e}. \quad (6.31)$$

Proof. Note that in the case where $\frac{\sigma_{ij}}{\sigma_{kl}} \to 1$ as $\sigma_{ij} \to \infty$, $i \neq j$, and $\sigma_{kl} \to \infty$, $k \neq l$, it follows that $\lim_{\sigma_{ij}\to\infty, i\neq j} A^{-1}$ is given by (6.22). Next, with $D = \mathrm{diag}[\sigma_{11}, ..., \sigma_{qq}]$ and $X = -qI_q + \mathbf{e}\mathbf{e}^{\mathrm{T}}$, it follows that $A = -D + \sigma X = -D^{\frac{1}{2}}(I_q - \sigma D^{-\frac{1}{2}}XD^{-\frac{1}{2}})D^{\frac{1}{2}}$. Now, it follows from Lemmas 6.2 and 6.3 that

$$E_\infty = \lim_{\sigma_{ij}\to\infty, i\neq j}(-A^{-1}S) = \frac{\mathbf{e}\mathbf{e}^{\mathrm{T}}}{\mathbf{e}^{\mathrm{T}}D\mathbf{e}}S = \left[\frac{\mathbf{e}^{\mathrm{T}}S}{\sum_{i=1}^{q}\sigma_{ii}}\right]\mathbf{e}, \quad (6.32)$$

which proves the result. $\square$

Proposition 6.4 shows that in the limit of strong coupling, the steady-state energy distribution E_∞ given by (6.17) becomes

$$E_\infty = \lim_{\sigma_{ij}\to\infty, i\neq j}(-A^{-1}S) = \left[\frac{\mathbf{e}^{\mathrm{T}}S}{\sum_{i=1}^{q}\sigma_{ii}}\right]\mathbf{e}, \quad (6.33)$$

which implies energy equipartition. This result first appeared in [10].

Chapter Seven

Continuum Thermodynamics

7.1 Conservation Laws in Continuum Thermodynamics

In this chapter we extend the results of Chapter 3 to the case of continuum thermodynamic systems, where the subsystems are uniformly distributed over an n-dimensional space. Since these thermodynamic systems involve distributed subsystems, they are described by partial differential equations and hence are infinite-dimensional systems. Our formulation in this chapter involves a unification of the behavior of heat as described by the equations of thermal transfer and classical thermodynamics. With the notable exception of [11], the amalgamation of these classical disciplines of physics is virtually nonexistent in the literature. Specifically, we consider continuous dynamical systems $\mathcal{G}$ defined over a compact connected set $\mathcal{V} \subset \mathbb{R}^n$ with a smooth (at least C^1) boundary $\partial\mathcal{V}$ and volume $\mathcal{V}_{\text{vol}}$. Furthermore, let $\mathcal{X}$ denote a space of two-times continuously differentiable scalar functions defined on $\mathcal{V}$, let $u(x, t)$, where $u : \mathcal{V} \times [0, \infty) \to \overline{\mathbb{R}}_+$, denote the energy density of the dynamical system $\mathcal{G}$ at the point $x \triangleq [x_1, ..., x_n]^{\mathrm{T}} \in \mathcal{V}$ and time instant $t \geq t_0$, let $\phi : \mathcal{V} \times \overline{\mathbb{R}}_+ \times \mathbb{R}^n \to \mathbb{R}^n$ denote the system energy flow within the continuum $\mathcal{V}$, that is, $\phi(x, u(x, t), \nabla u(x, t)) = [\phi_1(x, u(x, t), \nabla u(x, t)), ..., \phi_n(x, u(x, t), \nabla u(x, t))]^{\mathrm{T}}$, where $\phi_i(\cdot, \cdot, \cdot)$ denotes the energy flow through a unit area per unit time in the x_i direction for all $i = 1, ..., n$, and $\nabla u(x, t) \triangleq [D_1 u(x, t), ..., D_n u(x, t)]$, $x \in \mathcal{V}$, $t \geq t_0$, denotes the gradient of $u(\cdot, t)$ with respect to the spatial variable x, and let $s : \mathcal{V} \times [0, \infty) \to \overline{\mathbb{R}}_+$ denote the energy (heat) flow into a unit volume per unit time from sources uniformly distributed over $\mathcal{V}$.

To obtain the power balance equation for a uniformly distributed thermodynamic system, note that for any smooth, bounded region $\mathcal{V} \subset \mathbb{R}^n$, the integral $\int_{\mathcal{V}} u(x, t) d\mathcal{V}$ denotes the total amount of energy within $\mathcal{V}$ at time t. Hence, the rate of change of energy within $\mathcal{V}$ is governed by the *flux* function $\phi : \mathcal{V} \times \overline{\mathbb{R}}_+ \times \mathbb{R}^n \to \mathbb{R}^n$ and the external supplied power $s : \mathcal{V} \times [0, \infty) \to \overline{\mathbb{R}}_+$, which control the rate of loss and increase of the total energy through the boundary $\partial\mathcal{V}$ and the interior $\overset{\circ}{\mathcal{V}}$ of $\mathcal{V}$, respectively. Hence, for each time t,

$$\frac{\mathrm{d}}{\mathrm{d}t} \int_{\mathcal{V}} u(x,t)\mathrm{d}\mathcal{V} = -\int_{\partial\mathcal{V}} \phi(x, u(x,t), \nabla u(x,t)) \cdot \hat{n}(x)\mathrm{d}\mathcal{S}_{\mathcal{V}}$$
$$+ \int_{\mathcal{V}} s(x,t)\mathrm{d}\mathcal{V}, \tag{7.1}$$

where $\hat{n}(x)$ denotes the outward normal vector to the boundary $\partial\mathcal{V}$ (at x) of the set $\mathcal{V}$, $\mathrm{d}\mathcal{S}_{\mathcal{V}}$ denotes an infinitesimal surface element of the boundary of the set $\mathcal{V}$, and "$\cdot$" denotes the dot product in $\mathbb{R}^n$.

Using the divergence theorem, it follows from (7.1) that

$$\frac{\mathrm{d}}{\mathrm{d}t} \int_{\mathcal{V}} u(x,t)\mathrm{d}\mathcal{V} = -\int_{\partial\mathcal{V}} \phi(x, u(x,t), \nabla u(x,t)) \cdot \hat{n}(x)\mathrm{d}\mathcal{S}_{\mathcal{V}}$$
$$+ \int_{\mathcal{V}} s(x,t)\mathrm{d}\mathcal{V}$$
$$= -\int_{\mathcal{V}} \nabla \cdot \phi(x, u(x,t), \nabla u(x,t))\mathrm{d}\mathcal{V} + \int_{\mathcal{V}} s(x,t)\mathrm{d}\mathcal{V}, \tag{7.2}$$

where ∇ denotes the nabla operator. Since the region $\mathcal{V} \subset \mathbb{R}^n$ is arbitrary, it follows that the power balance equation over a unit volume within the continuum $\mathcal{V}$ involving the rate of energy density change, the external supplied power (heat flux), and the energy (heat) flow within the continuum is given by

$$\frac{\partial u(x,t)}{\partial t} = -\nabla \cdot \phi(x, u(x,t), \nabla u(x,t)) + s(x,t), \quad x \in \mathcal{V}, \quad t \geq t_0, \tag{7.3}$$
$$u(x,t_0) = u_{t_0}(x), \quad x \in \mathcal{V}, \quad \phi(x, u(x,t), \nabla u(x,t)) \cdot \hat{n}(x) \geq 0, \quad x \in \partial\mathcal{V}, \quad t \geq t_0, \tag{7.4}$$

where $u_{t_0} \in \mathcal{X}$ is a given initial energy density distribution.

The power balance (conservation) equation (7.3) describes the time evolution of the energy density $u(x,t)$ over the region $\mathcal{V}$, while the boundary condition in (7.4) involving the dot product implies that the energy of the system $\mathcal{G}$ can either be stored or dissipated but not supplied through the boundary of $\mathcal{V}$. Here, for simplicity of exposition, we assume that there is no work done by the system on the environment nor is there work done by the environment on the system. This extension can be easily handled by modifying the natural boundary condition for (7.3). In particular, this case would require that the system (7.3) and (7.4) is such that at every instant of time the domain $\mathcal{V}$ and its boundary $\partial\mathcal{V}$ are defined as $\mathcal{V} = \{x \in \mathbb{R}^n : f(x,t) \leq 0, t \geq t_0\}$

and $\partial \mathcal{V} = \{x \in \mathbb{R}^n : f(x,t) = 0, \, t \geq t_0\}$, where $f : \mathbb{R}^n \times [0,\infty) \to \mathbb{R}$ is a given continuously differentiable function, and consequently, the outward normal vector to the boundary $\partial \mathcal{V}$ at $x \in \partial \mathcal{V}$ and time $t \geq t_0$ is given by $\hat{n}^{\mathrm{T}}(x,t) = \nabla f(x,t)$.

We denote the energy density distribution over the set $\mathcal{V}$ at time $t \geq t_0$ by $u_t \in \mathcal{X}$ so that for each $t \geq t_0$ the set of mappings generated by $u_t(x) \equiv u(x,t)$ for every $x \in \mathcal{V}$ gives the *flow* of $\mathcal{G}$. We assume that the function $\phi(\cdot, \cdot, \cdot)$ is continuously differentiable so that (7.3) and (7.4) admits a unique solution $u(x,t)$, $x \in \mathcal{V}$, $t \geq t_0$, and $u(\cdot, t) \in \mathcal{X}$, $t \geq t_0$, is continuously dependent on the initial energy density distribution $u_{t_0}(x)$, $x \in \mathcal{V}$. It is well known, however, that nonlinear partial differential equations need not have smooth differentiable solutions (*classical solutions*), and one has to use the notion of distributions that provides a framework in which the energy density function $u(x,t)$ may be differentiated in a generalized sense infinitely often [38]. In this case, one has a well-defined notion of solutions that have jump discontinuities, which propagate as shock waves. Thus, one has to deal with *generalized* or *weak* solutions wherein uniqueness is lost. In this case, the *Clausius-Duhem inequality* is invoked for identifying the physically relevant (i.e., thermodynamically admissible) solution. (For further details, see [33, 38].) If u_{t_0} is a two-times continuously differentiable function with compact support and its derivative is sufficiently small on $[t_0, \infty)$, then the classical solution to (7.3) and (7.4) breaks down at a finite time. As a consequence of this, one may only hope to find generalized (or weak) solutions to (7.3) and (7.4) over the semi-infinite interval $[t_0, \infty)$, that is, $\mathcal{L}_\infty$ functions[1] $u(\cdot, \cdot)$ that satisfy (7.3) in the sense of distributions, which provides a framework in which $u(\cdot, \cdot)$ may be differentiated in a general sense infinitely often.

Next, we establish the uniqueness of the internal energy functional $U(u_t)$, $u_t \in \mathcal{X}$, for the dynamical system $\mathcal{G}$ defined by

$$U(u_{t_0}) \triangleq \int_{\mathcal{V}} u(x, t_0) \mathrm{d}\mathcal{V}, \quad u_{t_0} \in \mathcal{X}. \tag{7.5}$$

First, however, the following result on local controllability of our continuum thermodynamic model is required. For this result, let

---

[1] $\mathcal{L}_\infty$ denotes the space of bounded Lebesgue measurable functions on $\mathcal{V}$ and provides the broadest framework for weak solutions. Alternatively, a natural function class for weak solutions is the space $\mathcal{BV}$ consisting of functions of bounded variation. Recall that a bounded measurable function $u(x,t)$ has locally bounded variation if its distributional derivatives are locally finite Radon measures.

$\mathcal{L}_p = \mathcal{L}_p(\mathcal{V})$ denote a Lebesgue space, that is,

$$\mathcal{L}_p = \{u_t : \mathcal{V} \to \mathbb{R} : u_t \text{ is measurable}^2 \text{ and } \|u_t\|_{\mathcal{L}_p} < \infty\},$$

where

$$\|u_t\|_{\mathcal{L}_p} \triangleq \left[\int_{\mathcal{V}} |u_t|^p \mathrm{d}\mathcal{V}\right]^{1/p}, \quad 1 \le p < \infty, \tag{7.6}$$

and if $p = \infty$,

$$\|u_t\|_{\mathcal{L}_\infty} \triangleq \operatorname*{ess\,sup}_{\mathcal{V}} |u_t|, \tag{7.7}$$

where "ess" denotes essential.

Lemma 7.1 *Consider the dynamical system $\mathcal{G}$ with power balance equation (7.3) and (7.4). Then for every equilibrium state $u_e(\cdot) \in \mathcal{X}$ and every $\varepsilon > 0$ and $T > 0$, there exist $s_e : \mathcal{V} \to \mathbb{R}$, $\alpha > 0$, and $\hat{T} \in [0, T]$ such that for every $\hat{u}(\cdot) \in \mathcal{X}$ with $\|\hat{u} - u_e\|_{\mathcal{L}_p} \le \alpha T$, there exists $s : \mathcal{V} \times [0, \hat{T}] \to \mathbb{R}$ such that $\|s(\cdot, t) - s_e(\cdot)\|_{\mathcal{L}_p} \le \varepsilon$, $t \in [0, \hat{T}]$, and $u(x, t) = u_e(x) + \frac{\hat{u}(x) - u_e(x)}{\hat{T}} t$, $x \in \mathcal{V}$, $t \in [0, \hat{T}]$.*

Proof. Note that with $s_e(x) = \nabla \cdot \phi(x, u_e(x), \nabla u_e(x))$, $x \in \mathcal{V}$, the state $u_e(\cdot) \in \mathcal{X}$ is an equilibrium state of (7.3) and (7.4). Let $\theta > 0$ and $T > 0$, and define

$$M(\theta, T) \triangleq \sup_{u(\cdot) \in \overline{\mathcal{B}}_1(0), t \in [0,T]} \| -\nabla \cdot \phi(\cdot, u_e(\cdot) + \theta t u(\cdot), \nabla u_e(\cdot)$$
$$+ \theta t \nabla u(\cdot)) + s_e(\cdot)\|_{\mathcal{L}_p}, \tag{7.8}$$

where $\overline{\mathcal{B}}_1(0)$ denotes the closed unit ball in $\mathcal{L}_p$. Note that for every $T > 0$, $\lim_{\theta \to 0^+} M(\theta, T) = 0$, and for every $\theta > 0$, $\lim_{T \to 0^+} M(\theta, T) = 0$. Next, let $\varepsilon > 0$ and $T > 0$ be given, and let $\alpha > 0$ be such that $M(\alpha, T) + \alpha \le \varepsilon$. (The existence of such an α is guaranteed since $M(\alpha, T) \to 0$ as $\alpha \to 0^+$.) Now, let $\hat{u}(\cdot) \in \mathcal{X}$ be such that $\|\hat{u} - u_e\|_{\mathcal{L}_p} \le \alpha T$. With $\hat{T} \triangleq \frac{\|\hat{u} - u_e\|_{\mathcal{L}_p}}{\alpha} \le T$ and

$$s(x, t) = \nabla \cdot \phi(x, u(x, t), \nabla u(x, t)) + \alpha \frac{\hat{u}(x) - u_e(x)}{\|\hat{u} - u_e\|_{\mathcal{L}_p}}, \quad x \in \mathcal{V},$$
$$t \in [0, \hat{T}], \tag{7.9}$$

---

^{2}A function $u_t : \mathcal{V} \to \mathbb{R}$ is *measurable* if it is the pointwise limit, except for a set of Lebesgue measure zero, of a sequence of piecewise constant functions on $\mathcal{V}$.

it follows that

$$u(x,t) = u_e(x) + \frac{\hat{u}(x) - u_e(x)}{\|\hat{u} - u_e\|_{\mathcal{L}_p}}\alpha t, \quad x \in \mathcal{V}, \quad t \in [0, \hat{T}], \quad (7.10)$$

is a solution to (7.3) and (7.4). The result is now immediate by noting that $u(x, \hat{T}) = \hat{u}(x)$, $x \in \mathcal{V}$, and

$$\|s(\cdot, t) - s_e(\cdot)\|_{\mathcal{L}_p} \leq \| - \nabla \cdot \phi(\cdot, u(\cdot, t), \nabla u(\cdot, t)) + s_e(\cdot)\|_{\mathcal{L}_p} + \alpha$$
$$\leq M(\alpha, T) + \alpha$$
$$\leq \varepsilon, \quad t \in [0, \hat{T}]. \quad (7.11)$$

$\square$

It follows from Lemma 7.1 that the dynamical system $\mathcal{G}$ given by (7.3) and (7.4) is controllable to the zero energy density distribution, that is, for every $u_{t_0} \in \mathcal{X}$ there exists a finite time $t_f \geq t_0$ and $s(\cdot, \cdot) \in \mathcal{U}$ defined on $x \in \mathcal{V}$ and $t \in [t_0, t_f]$ such that the energy density distribution $u(x, t)$ can be driven from $u(x, t_0) = u_{t_0}(x)$ to $u(x, t_f) = 0$, $x \in \mathcal{V}$. Here, $\mathcal{U}$ denotes the set of all uniformly bounded in x and continuous in x and t energy inputs $s(\cdot, \cdot)$ to the system $\mathcal{G}$. In addition, it follows from Lemma 7.1 that (7.3) and (7.4) is reachable from the zero energy density distribution, that is, for every $u_{t_0} \in \mathcal{X}$ there exists a finite time $t_i \leq t_0$ and $s(\cdot, \cdot) \in \mathcal{U}$ defined on $x \in \mathcal{V}$ and $t \in [t_i, t_0]$ such that the energy density distribution $u(x, t)$ can be driven from $u(x, t_i) = 0$ to $u(x, t_0) = u_{t_0}(x)$, $x \in \mathcal{V}$. Next, let $\mathcal{U}_c \subset \mathcal{U}$ denote the set of all energy inputs to the system $\mathcal{G}$ such that for any $T \geq t_0$ the system energy distribution can be driven from $u(x, t_0) = u_{t_0}(x)$, $u_{t_0} \in \mathcal{X}$, to $u(x, T) = 0$, $x \in \mathcal{V}$, by $s(\cdot, \cdot) \in \mathcal{U}_c$, and we let $\mathcal{U}_r \subset \mathcal{U}$ denote the set of all energy inputs to the system $\mathcal{G}$ such that for any $T \geq -t_0$ the system energy distribution can be driven from $u(x, -T) = 0$ to $u(x, t_0) = u_{t_0}(x)$, $x \in \mathcal{V}$, $u_{t_0} \in \mathcal{X}$, by $s(\cdot, \cdot) \in \mathcal{U}_r$.

For the statement of the next result, define the available energy of the dynamical system $\mathcal{G}$ by

$$U_a(u_{t_0}) \triangleq - \inf_{s(\cdot, \cdot) \in \mathcal{U}, T \geq t_0} \int_{t_0}^{T} \left[\int_{\mathcal{V}} s(x, t) d\mathcal{V} \right.$$
$$\left. - \int_{\partial \mathcal{V}} \phi(x, u(x, t), \nabla u(x, t)) \cdot \hat{n}(x) d\mathcal{S}_{\mathcal{V}} \right] dt, \quad u_{t_0} \in \mathcal{X},$$
$$(7.12)$$

and the required energy supply of the dynamical system $\mathcal{G}$ by

$$
\begin{aligned}
U_{\mathrm{r}}(u_{t_0}) \triangleq \inf_{s(\cdot,\cdot) \in \mathcal{U}_{\mathrm{r}},\, T \geq -t_0} \int_{-T}^{t_0} & \left[\int_{\mathcal{V}} s(x,t) \mathrm{d}\mathcal{V} \right. \\
& \left. - \int_{\partial \mathcal{V}} \phi(x, u(x,t), \nabla u(x,t)) \cdot \hat{n}(x) \mathrm{d}\mathcal{S}_{\mathcal{V}} \right] \mathrm{d}t, \quad u_{t_0} \in \mathcal{X}.
\end{aligned}
$$
(7.13)

Note that the available energy $U_{\mathrm{a}}(u_{t_0})$ is the maximum amount of stored energy (net heat) that can be extracted from the dynamical system $\mathcal{G}$ at any time T, and the required energy supply $U_{\mathrm{r}}(u_{t_0})$ is the minimum amount of energy (net heat) that can be delivered to the dynamical system $\mathcal{G}$ to transfer it from a state of minimum energy density distribution $u_{-T} = u(x, -T) = 0$, $x \in \mathcal{V}$, to a state of given energy density distribution $u_{t_0} \in \mathcal{X}$.

Theorem 7.1 *Consider the dynamical system $\mathcal{G}$ with power balance equation (7.3) and (7.4). Then $\mathcal{G}$ is lossless with respect to the energy supply rate $\int_{\mathcal{V}} s(x,t) \mathrm{d}\mathcal{V} - \int_{\mathcal{V}} y(x,t) \mathrm{d}\mathcal{V}$, where $y(x,t) \equiv \nabla \cdot \phi(x, u(x,t), \nabla u(x,t))$, and with the unique energy storage functional corresponding to the total energy of the dynamical system $\mathcal{G}$ given by*

$$
\begin{aligned}
U(u_{t_0}) &= \int_{\mathcal{V}} u(x, t_0) \mathrm{d}\mathcal{V} \\
&= -\int_{t_0}^{T_+} \left[\int_{\mathcal{V}} s(x,t) \mathrm{d}\mathcal{V} - \int_{\mathcal{V}} y(x,t) \mathrm{d}\mathcal{V} \right] \mathrm{d}t \\
&= \int_{-T_-}^{t_0} \left[\int_{\mathcal{V}} s(x,t) \mathrm{d}\mathcal{V} - \int_{\mathcal{V}} y(x,t) \mathrm{d}\mathcal{V} \right] \mathrm{d}t, \quad u_{t_0} \in \mathcal{X}, \quad (7.14)
\end{aligned}
$$

where $u(x,t)$, $x \in \mathcal{V}$, $t \geq t_0$, is the solution to (7.3) and (7.4) with admissible input $s(\cdot,\cdot) \in \mathcal{U}$, $u_{-T_-} = 0$, $u_{T_+} = 0$, and $u_{t_0} \in \mathcal{X}$. Furthermore,

$$
0 \leq U_{\mathrm{a}}(u_{t_0}) = U(u_{t_0}) = U_{\mathrm{r}}(u_{t_0}) < \infty, \quad u_{t_0} \in \mathcal{X}. \quad (7.15)
$$

Proof. First, it follows from Lemma 7.1 that $\mathcal{G}$ is reachable from and controllable to the origin in $\mathcal{X}$. Next, it follows from (7.2) that

$$
\begin{aligned}
U(u_t) - U(u_{t_0}) = \int_{t_0}^{t} & \left[\int_{\mathcal{V}} s(x,t) \mathrm{d}\mathcal{V} \right. \\
& \left. - \int_{\mathcal{V}} \nabla \cdot \phi(x, u(x,t), \nabla u(x,t)) \mathrm{d}\mathcal{V} \right] \mathrm{d}t, \quad (7.16)
\end{aligned}
$$

which shows that the dynamical system $\mathcal{G}$ is lossless with respect to the energy supply rate $\int_{\mathcal{V}} s(x,t)d\mathcal{V} - \int_{\mathcal{V}} y(x,t)d\mathcal{V}$ and with the energy storage functional $U(u_{t_0}) = \int_{\mathcal{V}} u(x,t_0)d\mathcal{V}$, $u_{t_0} \in \mathcal{X}$. The remainder of the proof now follows identically as in the proof of Theorem 3.1 by noting that

$$\int_{\mathcal{V}} \nabla \cdot \phi(x, u(x,t), \nabla u(x,t))d\mathcal{V} = \int_{\partial\mathcal{V}} \phi(x, u(x,t), \nabla u(x,t)) \cdot \hat{n}(x)d\mathcal{S}_{\mathcal{V}},$$

(7.17)

which in turn follows from the divergence theorem.   □

It follows from (7.16) that the dynamical system $\mathcal{G}$ is lossless with respect to the net heat supply rate $\int_{\mathcal{V}} s(x,t)d\mathcal{V} - \int_{\mathcal{V}} y(x,t)d\mathcal{V}$ and with the unique energy storage functional $U(u_{t_0})$ given by (7.5). This is in essence a statement of the first law of thermodynamics for isochoric transformations of infinite-dimensional systems. As in Chapter 3, to ensure a thermodynamically consistent energy flow infinite-dimensional model, we require the following axioms which are analogous to Axioms i) and ii).

Axiom i)′ *For every $x \in \mathcal{V}$ and unit vector $\mathbf{u} \in \mathbb{R}^n$, $\phi(x, u_t(x), \nabla u_t(x)) \cdot \mathbf{u} = 0$ if and only if $\nabla u_t(x)\mathbf{u} = 0$.*

Axiom ii)′ *For every $x \in \mathcal{V}$ and unit vector $\mathbf{u} \in \mathbb{R}^n$, $\phi(x, u_t(x), \nabla u_t(x)) \cdot \mathbf{u} > 0$ if and only if $\nabla u_t(x)\mathbf{u} < 0$, and $\phi(x, u_t(x), \nabla u_t(x)) \cdot \mathbf{u} < 0$ if and only if $\nabla u_t(x)\mathbf{u} > 0$.*

Note that Axiom i)′ implies that $\phi_i(x, u_t(x), \nabla u_t(x)) = 0$ if and only if $D_i u_t(x) = 0$, $x \in \mathcal{V}$, $i = 1, ..., n$, while Axiom ii)′ implies that $\phi_i(x, u_t(x), \nabla u_t(x))D_i u_t(x) \leq 0$, $x \in \mathcal{V}$, $i = 1, ..., n$, which further implies that $\nabla u_t(x) \phi(x, u_t(x), \nabla u_t(x)) \leq 0$, $x \in \mathcal{V}$, that is, energy (heat) flows from regions of higher to lower energy densities. If $s(x,t) \equiv 0$, then Axioms i)′ and ii)′ along with the fact that $\phi(x, u(x,t), \nabla u(x,t)) \cdot \hat{n}(x) \geq 0$, $x \in \partial\mathcal{V}$, $t \geq t_0$, imply that at a given instant of time the energy of the dynamical system $\mathcal{G}$ can only be transported, stored, or dissipated but not created. We assume that if $u(\hat{x}, \hat{t}) = 0$ for some $\hat{x} \in \partial\mathcal{V}$ and $\hat{t} \geq t_0$, then $\phi(\hat{x}, u(\hat{x}, \hat{t}), \nabla u(\hat{x}, \hat{t})) = 0$, which, along with the boundary condition (7.4), implies that energy dissipation is not possible on the boundary of $\mathcal{V}$ through points with zero energy density.

With this assumption and Axiom ii)′ it follows that the solution $u(x,t)$, $x \in \mathcal{V}$, $t \geq t_0$, to (7.3) and (7.4) is nonnegative for all nonnegative initial energy density distributions $u_{t_0}(x) \geq 0$, $x \in \mathcal{V}$. To see this, note that if $u(\hat{x}, \hat{t}) = 0$ for some $\hat{x} \in \overset{\circ}{\mathcal{V}}$ and $\hat{t} \geq t_0$, then it follows from Axiom $ii)'$ that the energy flow $\phi(y, u(y, \hat{t}), \nabla u(y, \hat{t}))$ is directed towards the point $\hat{x}$ for all points y in a sufficiently small neighborhood of $\hat{x}$. This property along with the fact that $s : \mathcal{V} \times [0, \infty) \to \overline{\mathbb{R}}_+$ is a nonnegative function imply that $\frac{\partial u(\hat{x}, \hat{t})}{\partial t} \geq 0$. Alternatively, if $u(\hat{x}, \hat{t}) = 0$ for some $\hat{x} \in \partial \mathcal{V}$ and $\hat{t} \geq t_0$, then it follows from the above assumption that dissipation of energy through the point $\hat{x} \in \mathcal{V}$ is not possible, which, along with Axiom $ii)'$ and the fact that $s : \mathcal{V} \times [0, \infty) \to \overline{\mathbb{R}}_+$ is a nonnegative function, imply that $\frac{\partial u(\hat{x}, \hat{t})}{\partial t} \geq 0$ for $\hat{x} \in \partial \mathcal{V}$ and $\hat{t} \geq t_0$. Thus, the solution to (7.3) and (7.4) is nonnegative for all nonnegative initial energy density distributions. For the remainder of this chapter, $d\mathcal{V}$ represents an infinitesimal volume element of $\mathcal{V}$, $\mathcal{S}_\mathcal{V}$ denotes the surface enclosing $\mathcal{V}$, and $d\mathcal{S}_\mathcal{V}$ denotes an infinitesimal boundary element.

7.2 Entropy and Ectropy for Continuum Thermodynamics

In this section, we establish the classical Clausius inequality for our thermodynamically consistent infinite-dimensional energy flow model given by (7.3) and (7.4). For this result, note that it follows from Axiom $i)'$ that for the isolated dynamical system $\mathcal{G}$, that is, $s(x, t) \equiv 0$ and $\phi(x, u(x, t), \nabla u(x, t)) \cdot \hat{n}(x) \equiv 0$, the function $u(x, t) = \alpha$, $x \in \mathcal{V}$, $t \geq t_0$, $\alpha \geq 0$, is the solution to (7.3) and (7.4) with $u_{t_0}(x) = \alpha$, $x \in \mathcal{V}$. Thus, as in Chapter 3, we define an equilibrium process for the system $\mathcal{G}$ as a process where the trajectory of $\mathcal{G}$ moves along the equilibrium manifold $\mathcal{M}_e \triangleq \{u_t \in \mathcal{X} : u_t(x) = \alpha, x \in \mathcal{V}, \alpha \geq 0\}$, that is, $u(x, t) = \alpha(t), x \in \mathcal{V}, t \geq t_0$, for some $\mathcal{L}_\infty$ function $\alpha : [0, \infty) \to \overline{\mathbb{R}}_+$. A nonequilibrium process is a process that does not lie on $\mathcal{M}_e$. The next result establishes a Clausius-type inequality for equilibrium and nonequilibrium transformations of the infinite-dimensional dynamical system $\mathcal{G}$.

Proposition 7.1 *Consider the dynamical system $\mathcal{G}$ with power balance equation (7.3) and (7.4), and assume that Axioms $i)'$ and $ii)'$ hold. Then, for every initial energy density distribution $u_{t_0} \in \mathcal{X}$, $t_f \geq t_0$, and $s(\cdot, \cdot) \in \mathcal{U}$ such that $u_{t_f}(x) = u_{t_0}(x)$, $x \in \mathcal{V}$,*

$$\int_{t_0}^{t_f} \left[\int_{\mathcal{V}} \frac{s(x, t)}{c + u(x, t)} d\mathcal{V} - \int_{\partial \mathcal{V}} \frac{\phi(x, u(x, t), \nabla u(x, t)) \cdot \hat{n}(x)}{c + u(x, t)} d\mathcal{S}_\mathcal{V} \right] dt \leq 0,$$

$$(7.18)$$

where $c > 0$ and $u(x,t)$, $x \in \mathcal{V}$, $t \geq t_0$, is the solution to (7.3) and (7.4). Furthermore,

$$\int_{t_0}^{t_f} \left[\int_{\mathcal{V}} \frac{s(x,t)}{c + u(x,t)} d\mathcal{V} - \int_{\partial \mathcal{V}} \frac{\phi(x, u(x,t), \nabla u(x,t)) \cdot \hat{n}(x)}{c + u(x,t)} d\mathcal{S}_{\mathcal{V}} \right] dt = 0$$

(7.19)

if and only if there exists an $\mathcal{L}_\infty$ function $\alpha : [t_0, t_f] \to \overline{\mathbb{R}}_+$ such that $u(x,t) = \alpha(t)$, $x \in \mathcal{V}$, $t \in [t_0, t_f]$.

Proof. It follows from (7.3), the Green-Gauss theorem, and Axiom $ii)'$ that

$$\int_{t_0}^{t_f} \left[\int_{\mathcal{V}} \frac{s(x,t)}{c + u(x,t)} d\mathcal{V} - \int_{\partial \mathcal{V}} \frac{\phi(x, u(x,t), \nabla u(x,t)) \cdot \hat{n}(x)}{c + u(x,t)} d\mathcal{S}_{\mathcal{V}} \right] dt$$

$$= \int_{t_0}^{t_f} \int_{\mathcal{V}} \frac{\frac{\partial u(x,t)}{\partial t} + \nabla \cdot \phi(x, u(x,t), \nabla u(x,t))}{c + u(x,t)} d\mathcal{V} dt$$

$$- \int_{t_0}^{t_f} \int_{\partial \mathcal{V}} \frac{\phi(x, u(x,t), \nabla u(x,t)) \cdot \hat{n}(x)}{c + u(x,t)} d\mathcal{S}_{\mathcal{V}} dt$$

$$= \int_{\mathcal{V}} \log_e \left(\frac{c + u(x, t_f)}{c + u(x, t_0)} \right) d\mathcal{V}$$

$$+ \int_{t_0}^{t_f} \int_{\partial \mathcal{V}} \frac{\phi(x, u(x,t), \nabla u(x,t)) \cdot \hat{n}(x)}{c + u(x,t)} d\mathcal{S}_{\mathcal{V}} dt$$

$$+ \int_{t_0}^{t_f} \int_{\mathcal{V}} \frac{\nabla u(x,t) \phi(x, u(x,t), \nabla u(x,t))}{(c + u(x,t))^2} d\mathcal{V} dt$$

$$- \int_{t_0}^{t_f} \int_{\partial \mathcal{V}} \frac{\phi(x, u(x,t), \nabla u(x,t)) \cdot \hat{n}(x)}{c + u(x,t)} d\mathcal{S}_{\mathcal{V}} dt$$

$$= \int_{t_0}^{t_f} \int_{\mathcal{V}} \frac{\nabla u(x,t) \phi(x, u(x,t), \nabla u(x,t))}{(c + u(x,t))^2} d\mathcal{V} dt$$

$$\leq 0,$$

(7.20)

which proves (7.18).

To show (7.19), note that it follows from (7.20), Axiom $i)'$, and Axiom $ii)'$ that (7.19) holds if and only if $\nabla u(x,t) = 0$ for all $x \in \mathcal{V}$ and $t \in [t_0, t_f]$ or, equivalently, there exists an $\mathcal{L}_\infty$ function $\alpha : [t_0, t_f] \to \overline{\mathbb{R}}_+$ such that $u(x,t) = \alpha(t)$, $x \in \mathcal{V}$, $t \in [t_0, t_f]$. $\square$

Next, we define an entropy functional for the continuum dynamical system $\mathcal{G}$.

Definition 7.1 *For the dynamical system $\mathcal{G}$ with power balance equation (7.3) and (7.4), the functional $\mathcal{S} : \mathcal{X} \to \mathbb{R}$ satisfying*

$$\mathcal{S}(u_{t_2}) \geq \mathcal{S}(u_{t_1}) + \int_{t_1}^{t_2} q(t) \mathrm{d}t \tag{7.21}$$

for all $s(\cdot, \cdot) \in \mathcal{U}$ and $t_2 \geq t_1 \geq t_0$, where

$$q(t) \triangleq \int_{\mathcal{V}} \frac{s(x,t)}{c + u(x,t)} \mathrm{d}\mathcal{V} - \int_{\partial \mathcal{V}} \frac{\phi(x, u(x,t), \nabla u(x,t)) \cdot \hat{n}(x)}{c + u(x,t)} \mathrm{d}\mathcal{S}_{\mathcal{V}} \tag{7.22}$$

and $c > 0$, is called the entropy *functional of $\mathcal{G}$.*

In the next theorem, we show that (7.18) guarantees the existence of an entropy functional for the dynamical system $\mathcal{G}$ given by (7.3) and (7.4). For this result, define the available entropy of the dynamical system $\mathcal{G}$ by

$$\mathcal{S}_{\mathrm{a}}(u_{t_0}) \triangleq - \sup_{s(\cdot,\cdot) \in \mathcal{U}_{\mathrm{c}}, T \geq t_0} \int_{t_0}^{T} q(t) \mathrm{d}t, \tag{7.23}$$

where $q(t)$ is given by (7.22), $u(x, t_0) = u_{t_0}(x)$, $x \in \mathcal{V}$, $u_{t_0} \in \mathcal{X}$, and $u(x, T) = 0$, $x \in \mathcal{V}$, and define the required entropy supply of the dynamical system $\mathcal{G}$ by

$$\mathcal{S}_{\mathrm{r}}(u_{t_0}) \triangleq \sup_{s(\cdot,\cdot) \in \mathcal{U}_{\mathrm{r}}, T \geq -t_0} \int_{-T}^{t_0} q(t) \mathrm{d}t, \tag{7.24}$$

where $u(x, -T) = 0$, $x \in \mathcal{V}$, $u(x, t_0) = u_{t_0}(x)$, $x \in \mathcal{V}$, and $u_{t_0} \in \mathcal{X}$.

Theorem 7.2 *Consider the dynamical system $\mathcal{G}$ with power balance equation (7.3) and (7.4), and assume that Axiom ii)' holds. Then there exists an entropy functional for $\mathcal{G}$. Moreover, $\mathcal{S}_{\mathrm{a}}(u_{t_0})$, $u_{t_0} \in \mathcal{X}$, and $\mathcal{S}_{\mathrm{r}}(u_{t_0})$, $u_{t_0} \in \mathcal{X}$, are possible entropy functionals for $\mathcal{G}$ with $\mathcal{S}_{\mathrm{a}}(0) = \mathcal{S}_{\mathrm{r}}(0) = 0$. Finally, all entropy functionals $\mathcal{S}(u_{t_0})$, $u_{t_0} \in \mathcal{X}$, for $\mathcal{G}$ satisfy*

$$\mathcal{S}_{\mathrm{r}}(u_{t_0}) \leq \mathcal{S}(u_{t_0}) - \mathcal{S}(0) \leq \mathcal{S}_{\mathrm{a}}(u_{t_0}), \qquad u_{t_0} \in \mathcal{X}. \tag{7.25}$$

Proof. The proof is identical to the proof of Theorem 3.2. $\qquad \square$

The next result shows that all entropy functionals for $\mathcal{G}$ are continuous on $\mathcal{X}$ with norm $\| \cdot \|_{\mathcal{L}_1}$.

Theorem 7.3 *Consider the dynamical system $\mathcal{G}$ with power balance equation (7.3) and (7.4), and let $\mathcal{S} : \mathcal{X} \to \mathbb{R}$ be an entropy functional of $\mathcal{G}$. Then $\mathcal{S}(\cdot)$ is continuous on $\mathcal{X}$ with respect to the $\mathcal{L}_1$ norm.*

Proof. Let $u_e(\cdot) \in \mathcal{X}$ and $s_e : \mathcal{V} \to \mathbb{R}$ be such that $s_e(x) = \nabla \cdot \phi(x, u_e(x), \nabla u_e(x))$, $x \in \mathcal{V}$. Note that with $s(x, t) \equiv s_e(x)$, $x \in \mathcal{V}$, $u_e(\cdot)$ is an equilibrium state of the power balance equation (7.3) and (7.4). Next, it follows from Lemma 7.1 that for every $\varepsilon > 0$ and $T > 0$, there exist $s_e : \mathcal{V} \to \mathbb{R}$ and $\alpha > 0$ such that for every $\hat{u}(\cdot) \in \mathcal{X}$ with $\|\hat{u} - u_e\|_{\mathcal{L}_1} \leq \alpha T$, there exists $s : \mathcal{V} \times [0, \hat{T}] \to \mathbb{R}$ such that $\|s(\cdot, t) - s_e(\cdot)\|_{\mathcal{L}_1} \leq \varepsilon$, $t \in [0, \hat{T}]$, and $u(x, t) = u_e(x) + \frac{\hat{u}(x) - u_e(x)}{\hat{T}} t$, $x \in \mathcal{V}$, $t \in [0, \hat{T}]$, where $\hat{T} = \frac{\|\hat{u} - u_e\|_{\mathcal{L}_1}}{\alpha}$. Hence, for every $\delta > 0$ and $\varepsilon > 0$, there exist $s_e : \mathcal{V} \to \mathbb{R}$ and $\alpha > 0$ such that for every $\hat{u}(\cdot) \in \mathcal{X}$ with $\|\hat{u} - u_e\|_{\mathcal{L}_1} \leq \delta$, there exists $s : \mathcal{V} \times [0, \hat{T}] \to \mathbb{R}$ such that $\|s(\cdot, t) - s_e(\cdot)\|_{\mathcal{L}_1} \leq \varepsilon$, $t \in [0, \hat{T}]$, and $u(x, t) = u_e(x) + \frac{\hat{u}(x) - u_e(x)}{\hat{T}} t$, $x \in \mathcal{V}$, $t \in [0, \hat{T}]$, where $\hat{T} = \frac{\|\hat{u} - u_e\|_{\mathcal{L}_1}}{\alpha}$.

Next, since $\phi(\cdot, \cdot, \cdot)$ is continuous, it follows that there exists $M \in (0, \infty)$ such that

$$
\sup_{\|u - u_e\|_{\mathcal{L}_1} < \delta, \, \|s - s_e\|_{\mathcal{L}_1} < \varepsilon} \left\| \frac{s(x) - \nabla \cdot \phi(x, u(x), \nabla u(x))}{c + u(x)} + \frac{\nabla u(x) \phi(x, u(x), \nabla u(x))}{(c + u(x))^2} \right\|_{\mathcal{L}_1} = M. \quad (7.26)
$$

Hence, it follows that

$$
\left| \int_0^{\hat{T}} \left[\int_{\mathcal{V}} \frac{s(x, t) - \nabla \cdot \phi(x, u(x, t), \nabla u(x, t))}{c + u(x, t)} \mathrm{d}\mathcal{V} \right. \right.
$$
$$
\left. \left. + \int_{\mathcal{V}} \frac{\nabla u(x, t) \phi(x, u(x, t), \nabla u(x, t))}{(c + u(x, t))^2} \mathrm{d}\mathcal{V} \right] \mathrm{d}t \right|
$$
$$
\leq \int_0^{\hat{T}} \left\| \frac{s(x, t) - \nabla \cdot \phi(x, u(x, t), \nabla u(x, t))}{c + u(x, t)} \right.
$$
$$
\left. + \frac{\nabla u(x, t) \phi(x, u(x, t), \nabla u(x, t))}{(c + u(x, t))^2} \right\|_{\mathcal{L}_1} \mathrm{d}t
$$
$$
\leq M\hat{T}
$$
$$
= \frac{M}{\alpha} \|\hat{u} - u_e\|_{\mathcal{L}_1}. \quad (7.27)
$$

Next, if $\mathcal{S}(\cdot)$ is an entropy functional of $\mathcal{G}$, then, since $u(x, \hat{T}) = \hat{u}(x)$, $x \in \mathcal{V}$,

$$
\mathcal{S}(\hat{u}) \geq \mathcal{S}(u_e) + \int_0^{\hat{T}} \int_{\mathcal{V}} \frac{s(x, t)}{c + u(x, t)} \mathrm{d}\mathcal{V}\mathrm{d}t
$$

$$-\int_0^{\hat{T}}\int_{\partial\mathcal{V}}\frac{\phi(x,u(x,t),\nabla u(x,t))\cdot\hat{n}(x)}{c+u(x,t)}\mathrm{d}\mathcal{S}_{\mathcal{V}}\mathrm{d}t. \quad (7.28)$$

Hence, it follows from the Green-Gauss theorem that

$$\begin{aligned}
\mathcal{S}(u_\mathrm{e})-\mathcal{S}(\hat{u}) &\leq -\int_0^{\hat{T}}\int_{\mathcal{V}}\frac{s(x,t)}{c+u(x,t)}\mathrm{d}\mathcal{V}\mathrm{d}t \\
&\quad +\int_0^{\hat{T}}\int_{\partial\mathcal{V}}\frac{\phi(x,u(x,t),\nabla u(x,t))\cdot\hat{n}(x)}{c+u(x,t)}\mathrm{d}\mathcal{S}_{\mathcal{V}}\mathrm{d}t \\
&= -\int_0^{\hat{T}}\int_{\mathcal{V}}\frac{s(x,t)-\nabla\cdot\phi(x,u(x,t),\nabla u(x,t))}{c+u(x,t)}\mathrm{d}\mathcal{V}\mathrm{d}t \\
&\quad -\int_0^{\hat{T}}\int_{\mathcal{V}}\frac{\nabla u(x,t)\phi(x,u(x,t),\nabla u(x,t))}{(c+u(x,t))^2}\mathrm{d}\mathcal{V}\mathrm{d}t. \quad (7.29)
\end{aligned}$$

Now, if $\mathcal{S}(u_\mathrm{e})\geq\mathcal{S}(\hat{u})$, then combining (7.27) and (7.29) yields

$$|\mathcal{S}(u_\mathrm{e})-\mathcal{S}(\hat{u})|\leq\frac{M}{\alpha}\|\hat{u}-u_\mathrm{e}\|_{\mathcal{L}_1}. \quad (7.30)$$

Alternatively, if $\mathcal{S}(\hat{u})\geq\mathcal{S}(u_\mathrm{e})$, then (7.30) can be derived by reversing the roles of $u_\mathrm{e}(\cdot)$ and $\hat{u}(\cdot)$. Hence, it follows that $\mathcal{S}(\cdot)$ is continuous on $\mathcal{X}$ with norm $\|\cdot\|_{\mathcal{L}_1}$. $\qquad\square$

As for the finite-dimensional case, Definition 7.1 does not provide enough information to define the entropy uniquely for nonequilibrium continuum thermodynamics. Specifically, using a similar result to Proposition 3.3, it can be shown that all possible entropy functionals form a convex set, and hence, there exists a continuum of entropy functionals ranging from the required entropy supply $\mathcal{S}_\mathrm{r}(u_{t_0})$ to the available entropy $\mathcal{S}_\mathrm{a}(u_{t_0})$. The following two propositions address processes for equilibrium continuum thermodynamics wherein uniqueness is not an issue.

Proposition 7.2 *Consider the dynamical system $\mathcal{G}$ with power balance equation (7.3) and (7.4), and assume that Axioms i)' and ii)' hold. Then for every equilibrium energy density distribution $u_\mathrm{te}(x)=\alpha$, $x\in\mathcal{V}$, $\alpha\geq 0$, the entropy $\mathcal{S}(u_t)$, $u_t\in\mathcal{X}$, of $\mathcal{G}$ is unique (modulo a constant of integration) and is given by*

$$\mathcal{S}(u_t)-\mathcal{S}(0)=\mathcal{S}_\mathrm{a}(u_t)=\mathcal{S}_\mathrm{r}(u_t)=\mathcal{V}_\mathrm{vol}\log_e(c+\alpha)-\mathcal{V}_\mathrm{vol}\log_e c, \quad (7.31)$$

where $u_t(x)=u_\mathrm{te}(x)=\alpha$, $x\in\mathcal{V}$.

Proof. It follows from (7.3) and the Green-Gauss theorem that

$$
q(t) = \int_{\mathcal{V}} \frac{\frac{\partial u(x,t)}{\partial t} + \nabla \cdot \phi(x, u(x,t), \nabla u(x,t))}{c + u(x,t)} d\mathcal{V}
$$

$$
- \int_{\partial \mathcal{V}} \frac{\phi(x, u(x,t), \nabla u(x,t)) \cdot \hat{n}(x)}{c + u(x,t)} d\mathcal{S}_{\mathcal{V}}
$$

$$
= \int_{\mathcal{V}} \frac{1}{c + u(x,t)} \frac{\partial u(x,t)}{\partial t} d\mathcal{V}
$$

$$
+ \int_{\mathcal{V}} \frac{\nabla u(x,t) \phi(x, u(x,t), \nabla u(x,t))}{(c + u(x,t))^2} d\mathcal{V}. \qquad (7.32)
$$

Next, consider the entropy functional $\mathcal{S}_a(u_{t_0})$ given by (7.23), and let $u_{t_0}(x) = u_{te}(x) = \alpha$, $x \in \mathcal{V}$, $\alpha \geq 0$. Then it follows from (7.32) that

$$
\mathcal{S}_a(u_{te}) = - \sup_{s(\cdot,\cdot) \in \mathcal{U}_c, T \geq t_0} \left[\int_{t_0}^{T} \int_{\mathcal{V}} \frac{1}{c + u(x,t)} \frac{\partial u(x,t)}{\partial t} d\mathcal{V} dt \right.
$$

$$
\left. + \int_{t_0}^{T} \int_{\mathcal{V}} \frac{\nabla u(x,t) \phi(x, u(x,t), \nabla u(x,t))}{(c + u(x,t))^2} d\mathcal{V} dt \right]
$$

$$
= - \sup_{s(\cdot,\cdot) \in \mathcal{U}_c, T \geq t_0} \left[\int_{\mathcal{V}} \log_e \left(\frac{c}{c + \alpha} \right) d\mathcal{V} \right.
$$

$$
\left. + \int_{t_0}^{T} \int_{\mathcal{V}} \frac{\nabla u(x,t) \phi(x, u(x,t), \nabla u(x,t))}{(c + u(x,t))^2} d\mathcal{V} dt \right]
$$

$$
= \int_{\mathcal{V}} \log_e \left(\frac{c + \alpha}{c} \right) d\mathcal{V}
$$

$$
- \sup_{s(\cdot,\cdot) \in \mathcal{U}_c, T \geq t_0} \int_{t_0}^{T} \int_{\mathcal{V}} \frac{\nabla u(x,t) \phi(x, u(x,t), \nabla u(x,t))}{(c + u(x,t))^2} d\mathcal{V} dt.
$$

$$
(7.33)
$$

It follows from Axiom $ii)'$ that the supremum in (7.33) is taken over the set of negative semi-definite values. However, the zero value of the supremum is achieved on an equilibrium transformation for which $\phi(x, u(x,t), \nabla u(x,t)) \equiv 0$, and thus

$$
\mathcal{S}_a(u_{te}) = \mathcal{V}_{vol} \log_e(c + \alpha) - \mathcal{V}_{vol} \log_e c. \qquad (7.34)
$$

Similarly, it can be shown that $\mathcal{S}_r(u_{te}) = \mathcal{V}_{vol} \log_e(c + \alpha) - \mathcal{V}_{vol} \log_e c$. Finally, it follows from (7.25) that (7.31) holds. $\qquad \square$

Proposition 7.3 *Consider the dynamical system $\mathcal{G}$ with power balance equation (7.3) and (7.4), and assume that Axioms i)' and ii)' hold. Let $\mathcal{S}(\cdot)$ denote an entropy of $\mathcal{G}$, and let $u(x,t)$, $x \in \mathcal{V}$, $t \geq t_0$, be the solution to (7.3) and (7.4) with $u(x,t_0) = \alpha_0$, $x \in \mathcal{V}$, and $u(x,t_1) = \alpha_1$, $x \in \mathcal{V}$, where α_0, $\alpha_1 \geq 0$. Then*

$$\mathcal{S}(u_{t_1}) = \mathcal{S}(u_{t_0}) + \int_{t_0}^{t_1} q(t)\mathrm{d}t \qquad (7.35)$$

if and only if there exists an $\mathcal{L}_\infty$ function $\alpha : [t_0, t_1] \to \overline{\mathbb{R}}_+$ such that $\alpha(t_0) = \alpha_0$, $\alpha(t_1) = \alpha_1$, and $u(x,t) = \alpha(t)$, $x \in \mathcal{V}$, $t \in [t_0, t_1]$.

Proof. It follows from Proposition 7.2 that

$$\mathcal{S}(u_{t_1}) - \mathcal{S}(u_{t_0}) = \mathcal{V}_{\mathrm{vol}} \log_e(c + \alpha_1) - \mathcal{V}_{\mathrm{vol}} \log_e(c + \alpha_0). \qquad (7.36)$$

Furthermore, it follows from (7.32) that

$$\int_{t_0}^{t_1} q(t)\mathrm{d}t = \mathcal{V}_{\mathrm{vol}} \log_e\left(\frac{c + \alpha_1}{c + \alpha_0}\right)$$
$$+ \int_{t_0}^{t_1} \int_{\mathcal{V}} \frac{\nabla u(x,t)\phi(x, u(x,t), \nabla u(x,t))}{(c + u(x,t))^2} \mathrm{d}\mathcal{V}\mathrm{d}t. \qquad (7.37)$$

Now, it follows from Axioms i)' and ii)' that (7.35) holds if and only if $\nabla u(x,t) = 0$, $x \in \mathcal{V}$, $t \in [t_0, t_1]$, or, equivalently, there exists an $\mathcal{L}_\infty$ function $\alpha : [t_0, t_1] \to \overline{\mathbb{R}}_+$ such that $u(x,t) = \alpha(t)$, $x \in \mathcal{V}$, $t \in [t_0, t_1]$, $\alpha(t_0) = \alpha_0$, and $\alpha(t_1) = \alpha_1$. $\square$

In the next theorem, we present a unique, continuously differentiable entropy functional for the dynamical system $\mathcal{G}$. This result holds for equilibrium and nonequilibrium processes.

Theorem 7.4 *Consider the dynamical system $\mathcal{G}$ with power balance equation (7.3) and (7.4), and assume that Axioms i)' and ii)' hold. Then the functional $S : \mathcal{X} \to \mathbb{R}$ given by*

$$S(u_t) = \int_{\mathcal{V}} \log_e(c + u_t(x))\mathrm{d}\mathcal{V} - \mathcal{V}_{\mathrm{vol}} \log_e c \qquad (7.38)$$

is a unique (modulo a constant of integration), continuously differentiable entropy functional of $\mathcal{G}$. Furthermore, if $u_t \notin \mathcal{M}_\mathrm{e}$, $t \geq t_0$, where $u_t = u(x,t)$ denotes the solution to (7.3) and (7.4) and $\mathcal{M}_\mathrm{e} = \{u_t \in \mathcal{X} : u_t = \alpha,\ \alpha \geq 0\}$, then (7.38) satisfies

$$\mathcal{S}(u_{t_2}) > \mathcal{S}(u_{t_1}) + \int_{t_1}^{t_2} q(t)\mathrm{d}t. \qquad (7.39)$$

Proof. It follows from the Green-Gauss theorem, Axiom $ii)'$, and (7.38) that

$$
\begin{aligned}
\dot{\mathcal{S}}(u_t) &= \int_{\mathcal{V}} \frac{1}{c + u(x,t)} \frac{\partial u(x,t)}{\partial t} \, d\mathcal{V} \\
&= \int_{\mathcal{V}} \frac{1}{c + u(x,t)} \left(-\nabla \cdot \phi(x, u(x,t), \nabla u(x,t)) + s(x,t) \right) d\mathcal{V} \\
&= -\int_{\mathcal{V}} \frac{\nabla u(x,t) \phi(x, u(x,t), \nabla u(x,t))}{(c + u(x,t))^2} d\mathcal{V} \\
&\quad - \int_{\partial \mathcal{V}} \frac{\phi(x, u(x,t), \nabla u(x,t)) \cdot \hat{n}(x)}{c + u(x,t)} d\mathcal{S}_{\mathcal{V}} \\
&\quad + \int_{\mathcal{V}} \frac{s(x,t)}{c + u(x,t)} d\mathcal{V} \\
&\geq q(t).
\end{aligned}
\tag{7.40}
$$

Now, integrating (7.40) over $[t_1, t_2]$ yields (7.21). Furthermore, if $u_t \notin \mathcal{M}_e$, $t \geq t_0$, then it follows from Axiom $i)'$, Axiom $ii)'$, and (7.40) that (7.39) holds.

The uniqueness of the entropy functional (7.38) follows as in the proof of Theorem 3.4.   □

The next result shows that for every nontrivial trajectory of $\mathcal{G}$, the dynamical system $\mathcal{G}$ is state irreversible. For this result, let $\mathcal{W}_{[t_0,t_1]}$ denote the set of all possible energy density distributions of $\mathcal{G}$ over the time interval $[t_0, t_1]$ given by

$$
\mathcal{W}_{[t_0,t_1]} \triangleq \{ s^u : [t_0, t_1] \times \mathcal{U} \to \mathcal{X} : s^u(\cdot, s(\cdot, \cdot)) \text{ satisfies}
$$
$$
(7.3) \text{ and } (7.4) \}.
\tag{7.41}
$$

Theorem 7.5 *Consider the dynamical system $\mathcal{G}$ with power balance equation (7.3) and (7.4), and assume that Axioms $i)'$ and $ii)'$ hold. Furthermore, let $s^u(\cdot, s(\cdot, \cdot)) \in \mathcal{W}_{[t_0,t_1]}$, where $s(\cdot, \cdot) \in \mathcal{U}$. Then $s^u(\cdot, s(\cdot, \cdot))$ is an $I_{\mathcal{X}}$-reversible trajectory of $\mathcal{G}$ if and only if $s^u(t, s(x,t)) \in \mathcal{M}_e$, $t \in [t_0, t_1]$.*

Proof. The proof is similar to the proof of Theorem 3.5.   □

Next, we establish a dual inequality to inequality (7.18) that is satisfied for our thermodynamically consistent energy flow model.

Proposition 7.4 *Consider the dynamical system $\mathcal{G}$ with power balance equation (7.3) and (7.4), and assume that Axioms $i)'$ and $ii)'$*

hold. Then, for every initial energy density distribution $u_{t_0} \in \mathcal{X}$, $t_f \geq t_0$, and $s(\cdot, \cdot) \in \mathcal{U}$ such that $u_{t_f}(x) = u_{t_0}(x)$, $x \in \mathcal{V}$,

$$\int_{t_0}^{t_f} \int_{\mathcal{V}} u(x,t)s(x,t) d\mathcal{V} dt$$

$$- \int_{t_0}^{t_f} \int_{\partial\mathcal{V}} u(x,t)\phi(x, u(x,t), \nabla u(x,t)) \cdot \hat{n}(x) d\mathcal{S}_{\mathcal{V}} dt \geq 0, \quad (7.42)$$

where $u(x,t)$, $x \in \mathcal{V}$, $t \geq t_0$, is the solution to (7.3) and (7.4). Furthermore,

$$\int_{t_0}^{t_f} \int_{\mathcal{V}} u(x,t)s(x,t) d\mathcal{V} dt$$

$$- \int_{t_0}^{t_f} \int_{\partial\mathcal{V}} u(x,t)\phi(x, u(x,t), \nabla u(x,t)) \cdot \hat{n}(x) d\mathcal{S}_{\mathcal{V}} dt = 0 \quad (7.43)$$

if and only if there exists an $\mathcal{L}_\infty$ function $\alpha : [t_0, t_f] \to \overline{\mathbb{R}}_+$ such that $u(x,t) = \alpha(t)$, $x \in \mathcal{V}$, $t \in [t_0, t_f]$.

Proof. It follows from (7.3), the Green-Gauss theorem, and Axiom $ii)'$ that

$$\int_{t_0}^{t_f} \int_{\mathcal{V}} u(x,t)s(x,t) d\mathcal{V} dt$$

$$- \int_{t_0}^{t_f} \int_{\partial\mathcal{V}} u(x,t)\phi(x, u(x,t), \nabla u(x,t)) \cdot \hat{n}(x) d\mathcal{S}_{\mathcal{V}} dt$$

$$= \int_{t_0}^{t_f} \int_{\mathcal{V}} u(x,t) \left(\frac{\partial u(x,t)}{\partial t} + \nabla \cdot \phi(x, u(x,t), \nabla u(x,t)) \right) d\mathcal{V} dt$$

$$- \int_{t_0}^{t_f} \int_{\partial\mathcal{V}} u(x,t)\phi(x, u(x,t), \nabla u(x,t)) \cdot \hat{n}(x) d\mathcal{S}_{\mathcal{V}} dt$$

$$= \int_{\mathcal{V}} \left[\tfrac{1}{2} u^2(x, t_f) - \tfrac{1}{2} u^2(x, t_0) \right] d\mathcal{V}$$

$$+ \int_{t_0}^{t_f} \int_{\partial\mathcal{V}} u(x,t)\phi(x, u(x,t), \nabla u(x,t)) \cdot \hat{n}(x) d\mathcal{S}_{\mathcal{V}} dt$$

$$- \int_{t_0}^{t_f} \int_{\mathcal{V}} \nabla u(x,t)\phi(x, u(x,t), \nabla u(x,t)) d\mathcal{V} dt$$

$$- \int_{t_0}^{t_f} \int_{\partial\mathcal{V}} u(x,t)\phi(x, u(x,t), \nabla u(x,t)) \cdot \hat{n}(x) d\mathcal{S}_{\mathcal{V}} dt$$

$$= - \int_{t_0}^{t_f} \int_{\mathcal{V}} \nabla u(x,t)\phi(x, u(x,t), \nabla u(x,t)) d\mathcal{V} dt$$

$$\geq 0, \tag{7.44}$$

which proves (7.42).

To show (7.43), note that it follows from (7.44), Axiom $i)'$, and Axiom $ii)'$ that (7.43) holds if and only if $\nabla u(x,t) = 0$ for all $x \in \mathcal{V}$ and $t \in [t_0, t_f]$ or, equivalently, there exists an $\mathcal{L}_\infty$ function $\alpha : [t_0, t_f] \to \overline{\mathbb{R}}_+$ such that $u(x,t) = \alpha(t)$, $x \in \mathcal{V}$, $t \in [t_0, t_f]$. $\qquad\square$

Definition 7.2 *For the dynamical system $\mathcal{G}$ with power balance equation (7.3) and (7.4), the functional $\mathcal{E} : \mathcal{X} \to \mathbb{R}$ satisfying*

$$\mathcal{E}(u_{t_2}) \leq \mathcal{E}(u_{t_1}) + \mathcal{V}_{\mathrm{vol}} \int_{t_1}^{t_2} \hat{q}(t)\mathrm{d}t \tag{7.45}$$

for all $s(\cdot, \cdot) \in \mathcal{U}$ and $t_2 \geq t_1 \geq t_0$, where

$$\hat{q}(t) \triangleq \int_{\mathcal{V}} u(x,t)s(x,t)\mathrm{d}\mathcal{V}$$
$$- \int_{\partial\mathcal{V}} u(x,t)\phi(x, u(x,t), \nabla u(x,t)) \cdot \hat{n}(x)\mathrm{d}\mathcal{S}_{\mathcal{V}}, \tag{7.46}$$

is called the ectropy *functional of $\mathcal{G}$.*

The next theorem shows that (7.42) guarantees the existence of an ectropy functional for the dynamical system $\mathcal{G}$ given by (7.3) and (7.4). For this result, define the available ectropy of the dynamical system $\mathcal{G}$ by

$$\mathcal{E}_{\mathrm{a}}(u_{t_0}) \triangleq -\mathcal{V}_{\mathrm{vol}} \inf_{s(\cdot,\cdot)\in\mathcal{U}_{\mathrm{c}},\, T\geq t_0} \int_{t_0}^{T} \hat{q}(t)\mathrm{d}t, \tag{7.47}$$

where $\hat{q}(t)$ is given by (7.46), $u(x, t_0) = u_{t_0}(x)$, $x \in \mathcal{V}$, $u_{t_0} \in \mathcal{X}$, and $u(x, T) = 0$, $x \in \mathcal{V}$, and define the required ectropy supply of the dynamical system $\mathcal{G}$ by

$$\mathcal{E}_{\mathrm{r}}(u_{t_0}) \triangleq \mathcal{V}_{\mathrm{vol}} \inf_{s(\cdot,\cdot)\in\mathcal{U}_{\mathrm{r}},\, T\geq -t_0} \int_{-T}^{t_0} \hat{q}(t)\mathrm{d}t, \tag{7.48}$$

where $u(x, -T) = 0$, $x \in \mathcal{V}$, $u(x, t_0) = u_{t_0}(x)$, $x \in \mathcal{V}$, and $u_{t_0} \in \mathcal{X}$.

Theorem 7.6 *Consider the dynamical system $\mathcal{G}$ with power balance equation (7.3) and (7.4), and assume that Axiom $ii)'$ holds. Then there exists an ectropy functional for $\mathcal{G}$. Moreover, $\mathcal{E}_{\mathrm{a}}(u_{t_0})$, $u_{t_0} \in \mathcal{X}$, and $\mathcal{E}_{\mathrm{r}}(u_{t_0})$, $u_{t_0} \in \mathcal{X}$, are possible ectropy functionals for $\mathcal{G}$ with*

$\mathcal{E}_a(0) = \mathcal{E}_r(0) = 0$. *Finally, all ectropy functionals* $\mathcal{E}(u_{t_0})$, $u_{t_0} \in \mathcal{X}$, *for* $\mathcal{G}$ *satisfy*

$$\mathcal{E}_a(u_{t_0}) \leq \mathcal{E}(u_{t_0}) - \mathcal{E}(0) \leq \mathcal{E}_r(u_{t_0}), \quad u_{t_0} \in \mathcal{X}. \qquad (7.49)$$

Proof. The proof is identical to the proof of Theorem 3.6. $\qquad \square$

The next theorem shows that all ectropy functionals for $\mathcal{G}$ are continuous on $\mathcal{X}$ with norm $\| \cdot \|_{\mathcal{L}_1}$.

Theorem 7.7 *Consider the dynamical system* $\mathcal{G}$ *with power balance equation (7.3) and (7.4), and let* $\mathcal{E} : \mathcal{X} \to \mathbb{R}$ *be an ectropy functional of* $\mathcal{G}$. *Then* $\mathcal{E}(\cdot)$ *is continuous on* $\mathcal{X}$ *with respect to the* $\mathcal{L}_1$ *norm.*

Proof. The proof is identical to the proof of Theorem 7.3. $\qquad \square$

The following two propositions are dual to Propositions 7.2 and 7.3 and address equilibrium processes for continuum thermodynamics using ectropy notions.

Proposition 7.5 *Consider the dynamical system* $\mathcal{G}$ *with power balance equation (7.3) and (7.4), and assume that Axioms i)$'$ and ii)$'$ hold. Then for every energy density distribution* $u_{te}(x) = \alpha$, $x \in \mathcal{V}$, $\alpha \geq 0$, *the ectropy* $\mathcal{E}(u_t)$, $u_t \in \mathcal{X}$, *of* $\mathcal{G}$ *is unique (modulo a constant of integration) and is given by*

$$\mathcal{E}(u_t) - \mathcal{E}(0) = \mathcal{E}_a(u_t) = \mathcal{E}_r(u_t) = \frac{(\alpha \mathcal{V}_{\text{vol}})^2}{2}, \qquad (7.50)$$

where $u_t(x) = u_{te}(x) = \alpha$, $x \in \mathcal{V}$.

Proof. The proof is identical to the proof of Proposition 7.2. $\qquad \square$

Proposition 7.6 *Consider the dynamical system* $\mathcal{G}$ *with power balance equation (7.3) and (7.4), and assume that Axioms i)$'$ and ii)$'$ hold. Let* $\mathcal{E}(\cdot)$ *denote an ectropy of* $\mathcal{G}$, *and let* $u(x, t)$, $x \in \mathcal{V}$, $t \geq t_0$, *be the solution to (7.3) and (7.4) with* $u(x, t_0) = \alpha_0$, $x \in \mathcal{V}$, *and* $u(x, t_1) = \alpha_1$, $x \in \mathcal{V}$, *where* α_0, $\alpha_1 \geq 0$. *Then*

$$\mathcal{E}(u_{t_1}) = \mathcal{E}(u_{t_0}) + \mathcal{V}_{\text{vol}} \int_{t_0}^{t_1} \hat{q}(t) \mathrm{d}t \qquad (7.51)$$

if and only if there exists an $\mathcal{L}_\infty$ *function* $\alpha : [t_0, t_1] :\to \overline{\mathbb{R}}_+$ *such that* $\alpha(t_0) = \alpha_0$, $\alpha(t_1) = \alpha_1$, *and* $u(x, t) = \alpha(t)$, $x \in \mathcal{V}$, $t \in [t_0, t_1]$.

Proof. The proof is identical to the proof of Proposition 7.3. $\qquad$ $\square$

In the next theorem, we present a unique, continuously differentiable ectropy functional for the dynamical system $\mathcal{G}$. This result holds for equilibrium and nonequilibrium processes.

Theorem 7.8 *Consider the dynamical system $\mathcal{G}$ with power balance equation (7.3) and (7.4), and assume that Axioms i)' and ii)' hold. Then the functional $\mathcal{E} : \mathcal{X} \to \mathbb{R}$ given by*

$$\mathcal{E}(u_t) = \frac{\mathcal{V}_{\text{vol}}}{2} \int_{\mathcal{V}} u_t^2(x) \mathrm{d}\mathcal{V} \qquad (7.52)$$

is a unique (modulo a constant of integration), continuously differentiable ectropy functional of $\mathcal{G}$. Furthermore, if $u_t \notin \mathcal{M}_e$, $t \geq t_0$, where $u_t = u(x, t)$ denotes the solution to (7.3) and (7.4) and $\mathcal{M}_e = \{u_t \in \mathcal{X} : u_t = \alpha, \ \alpha \geq 0\}$, then (7.52) satisfies

$$\mathcal{E}(u_{t_2}) < \mathcal{E}(u_{t_1}) + \mathcal{V}_{\text{vol}} \int_{t_1}^{t_2} \hat{q}(t) \mathrm{d}t. \qquad (7.53)$$

Proof. It follows from the Green-Gauss theorem, Axiom $ii)'$, (7.3), and (7.52) that

$$\dot{\mathcal{E}}(u_t) = \mathcal{V}_{\text{vol}} \int_{\mathcal{V}} u(x, t) \frac{\partial u(x, t)}{\partial t} \mathrm{d}\mathcal{V}$$

$$= \mathcal{V}_{\text{vol}} \int_{\mathcal{V}} u(x, t) \left(-\nabla \cdot \phi(x, u(x, t), \nabla u(x, t)) + s(x, t) \right) \mathrm{d}\mathcal{V}$$

$$= -\mathcal{V}_{\text{vol}} \int_{\partial \mathcal{V}} u(x, t) \phi(x, u(x, t), \nabla u(x, t)) \cdot \hat{n}(x) \mathrm{d}\mathcal{S}_{\mathcal{V}}$$

$$\quad + \mathcal{V}_{\text{vol}} \int_{\mathcal{V}} \nabla u(x, t) \phi(x, u(x, t), \nabla u(x, t)) \mathrm{d}\mathcal{V}$$

$$\quad + \mathcal{V}_{\text{vol}} \int_{\mathcal{V}} u(x, t) s(x, t) \mathrm{d}\mathcal{V}$$

$$\leq \mathcal{V}_{\text{vol}} \hat{q}(t). \qquad (7.54)$$

Now, integrating (7.54) over $[t_1, t_2]$ yields (7.45). Furthermore, if $u_t \notin \mathcal{M}_e$, $t \geq t_0$, then it follows from Axiom $i)'$, Axiom $ii)'$, and (7.54) that (7.53) holds.

The uniqueness of the ectropy functional (7.52) follows as in the proof of Theorem 3.8. $\qquad$ $\square$

Inequality (7.21) is a generalization of Clausius' inequality for equilibrium and nonequilibrium thermodynamics as applied to infinite-dimensional systems, while inequality (7.45) is an anti–Clausius inequality that shows that a thermodynamically consistent infinite-dimensional dynamical system is dissipative with respect to the supply rate $\mathcal{V}_{\text{vol}}\hat{q}(t)$ and with storage functional corresponding to the system ectropy. In addition, note that it follows from (7.21) that the infinitesimal increment in the entropy of $\mathcal{G}$ over the infinitesimal time interval dt satisfies

$$dS(u_t) \geq \left[\int_{\mathcal{V}} \frac{s(x,t)}{c + u(x,t)} d\mathcal{V} \right.$$
$$\left. - \int_{\partial\mathcal{V}} \frac{\phi(x, u(x,t), \nabla u(x,t)) \cdot \hat{n}(x)}{c + u(x,t)} d\mathcal{S}_{\mathcal{V}} \right] dt, \quad (7.55)$$

where the shifted energy density $c + u(x,t)$ plays the role of absolute temperature at the spatial coordinate x and time t. For an isolated dynamical system $\mathcal{G}$ (that is, $s(x,t) \equiv 0$ and $\phi(x, u(x,t), \nabla u(x,t)) \cdot \hat{n}(x) \equiv 0$, $x \in \partial\mathcal{V}$), (7.21) and (7.45) yield the fundamental inequalities

$$S(u_{t_2}) \geq S(u_{t_1}), \quad t_2 \geq t_1, \quad (7.56)$$

and

$$\mathcal{E}(u_{t_2}) \leq \mathcal{E}(u_{t_1}), \quad t_2 \geq t_1. \quad (7.57)$$

Hence, for an isolated infinite-dimensional system $\mathcal{G}$, the entropy increases if and only if the ectropy decreases. It is important to note that (7.57) also holds in the case where $\phi(x, u(x,t), \nabla u(x,t)) \cdot \hat{n}(x) \not\equiv 0$, $x \in \partial\mathcal{V}$, whereas (7.56) does not necessarily hold in that case.

7.3 Semistability and Energy Equipartition in Continuum Thermodynamics

In this section, we show that the infinite-dimensional thermodynamic energy flow model has convergent flows to Lyapunov stable uniform equilibrium energy density distributions determined by the system initial energy density distribution. However, since our continuous dynamical system $\mathcal{G}$ is defined on the infinite-dimensional space $\mathcal{X}$, bounded orbits of $\mathcal{G}$ may not lie in a compact subset of $\mathcal{X}$, which is crucial to being able to invoke the invariance principle for infinite-dimensional dynamical systems [49]. This is in contrast to the dynamical system $\mathcal{G}$ considered in the previous chapters arising from a power balance (ordinary differential) equation defined on a finite-dimensional space $\mathbb{R}_+^q$, wherein local boundedness of an orbit of $\mathcal{G}$ ensures that the orbit belongs to a compact subset of $\overline{\mathbb{R}}_+^q$. Hence, to ensure that bounded orbits of $\mathcal{G}$ lie in compact sets, we construct a larger space $\mathcal{H}$ as a Sobolev space so that $\mathcal{X} \subset \mathcal{H}$, and by the Sobolev embedding theorem [92, 98], there exists a Banach space $\mathcal{B} \supset \mathcal{H}$ such that the unit ball in $\mathcal{H}$ belongs to a compact set in $\mathcal{B}$, that is, $\mathcal{H}$ is *compactly embedded* in $\mathcal{B}$. In this case, it follows from Proposition 2.2 that a bounded orbit of the dynamical system $\mathcal{G}$ defined on $\mathcal{H}$ has a nonempty, compact, connected invariant omega limit set in $\mathcal{B}$.

For the next result, $\mathcal{L}_2$ denotes the space of square-integrable Lebesgue measurable functions on $\mathcal{V}$ and the $\mathcal{L}_2$ operator norm $\|\cdot\|_{\mathcal{L}_2}$ on $\mathcal{X}$ is used for the definitions of Lyapunov, semi-, and asymptotic stability. Furthermore, we introduce the Sobolev spaces

$$\mathcal{W}_2^1(\mathcal{V}) \triangleq \{u_t : \mathcal{V} \to \mathbb{R} : u_t \in \mathrm{C}^1(\mathcal{V}) \cap \mathcal{L}_2(\mathcal{V}), (\nabla u_t)^{\mathrm{T}} \in \mathcal{L}_2(\mathcal{V})\}_{\mathrm{co}} \tag{7.58}$$

and

$$\mathcal{W}_2^0(\mathcal{V}) \triangleq \{u_t : \mathcal{V} \to \mathbb{R} : u_t \in \mathrm{C}^0(\mathcal{V}) \cap \mathcal{L}_2(\mathcal{V})\}_{\mathrm{co}} \subset \mathcal{L}_2(\mathcal{V}), \tag{7.59}$$

where $\mathrm{C}^r(\mathcal{V})$ denotes a function space defined on $\mathcal{V}$ with r-continuous derivatives and $\{\cdot\}_{\mathrm{co}}$ denotes completion[3] of $\{\cdot\}$ in $\mathcal{L}_2$ in the sense of [98], with norms

$$\|u_t\|_{\mathcal{W}_2^1} \triangleq \left[\int_{\mathcal{V}} \left(u_t^2(x) + \nabla u_t(x)\,(\nabla u_t(x))^{\mathrm{T}}\right) \mathrm{d}\mathcal{V}\right]^{\frac{1}{2}}, \tag{7.60}$$

$$\|u_t\|_{\mathcal{W}_2^0} \triangleq \|u_t\|_{\mathcal{L}_2} = \left[\int_{\mathcal{V}} u_t^2(x)\mathrm{d}\mathcal{V}\right]^{\frac{1}{2}}, \tag{7.61}$$

defined on $\mathcal{W}_2^1(\mathcal{V})$ and $\mathcal{W}_2^0(\mathcal{V})$, respectively, where the gradient $\nabla u_t(x)$ in (7.60) is interpreted in the sense of a generalized gradient [98]. Note that since the solutions to (7.3) and (7.4) are assumed to be two-times continuously differentiable functions on a compact set $\mathcal{V}$, it follows that $u_t(x)$, $t \geq t_0$, belongs to both $\mathcal{W}_2^1(\mathcal{V})$ and $\mathcal{W}_2^0(\mathcal{V})$.

---

[3]The space $\{\cdot\}$ defined as part of (7.58) is not complete with respect to the norm generated by the inner product (7.60). This space can be completed by adding the limit points of all Cauchy sequences in $\{\cdot\}$. In this way, $\{\cdot\}$ is embedded in the larger normed space $\{\cdot\}_{\mathrm{co}}$, which is complete. Of course, it follows from the Riesz-Fischer theorem [88, p. 125] that $\mathcal{L}_2$ is complete with respect to the norm generated by the inner product (7.61).

Theorem 7.9 *Consider the dynamical system $\mathcal{G}$ with power balance equation (7.3) and (7.4) with $s(x,t) \equiv 0$ and $\phi(x, u(x,t), \nabla u(x,t)) \cdot \hat{n}(x) \equiv 0$, $x \in \partial \mathcal{V}$. Assume that Axioms i)′ and ii)′ hold, and*

$$\nabla^2 u_t(x) \nabla \cdot \phi(x, u_t(x), \nabla u_t(x)) \leq 0, \quad x \in \mathcal{V}, \quad u_t \in \mathcal{W}_2^1(\mathcal{V}), \quad (7.62)$$

where $\nabla^2 \triangleq \nabla \cdot \nabla$ denotes the Laplacian operator. Then for every $\alpha \geq 0$, $u(x,t) \equiv \alpha$ is a semistable equilibrium state of (7.3) and (7.4). Furthermore, $u(x,t) \to \frac{1}{\mathcal{V}_{\mathrm{vol}}} \int_{\mathcal{V}} u_{t_0}(x) \mathrm{d}\mathcal{V}$ as $t \to \infty$ for every initial energy density distribution $u_{t_0} \in \mathcal{W}_2^1(\mathcal{V})$ and every $x \in \mathcal{V}$; moreover, $\frac{1}{\mathcal{V}_{\mathrm{vol}}} \int_{\mathcal{V}} u_{t_0}(x) \mathrm{d}\mathcal{V}$ is a semistable equilibrium distribution state of (7.3) and (7.4). Finally, if $s(x,t) \equiv 0$ and there exists at least one point $x_{\mathrm{p}} \in \partial \mathcal{V}$ such that $\phi(x_{\mathrm{p}}, u_t(x_{\mathrm{p}}), \nabla u_t(x_{\mathrm{p}})) \cdot \hat{n}(x_{\mathrm{p}}) > 0$ and $\phi(x_{\mathrm{p}}, u_t(x_{\mathrm{p}}), \nabla u_t(x_{\mathrm{p}})) \cdot \hat{n}(x_{\mathrm{p}}) = 0$ if and only if $u_t(x_{\mathrm{p}}) = 0$, then the zero solution $u(x,t) \equiv 0$ to (7.3) and (7.4) is a globally asymptotically stable equilibrium state of (7.3) and (7.4).

Proof. It follows from Axiom i)′ that $u(x,t) \equiv \alpha$, $\alpha \geq 0$, is an equilibrium state for (7.3), (7.4) with $s(x,t) \equiv 0$ and $\phi(x, u(x,t), \nabla u(x,t)) \cdot \hat{n}(x) \equiv 0$. To show Lyapunov stability of the equilibrium state $u(x,t) \equiv \alpha$, consider the shifted-system scaled ectropy $\mathcal{E}_{\mathrm{s}}(u_t) = \frac{1}{2} \int_{\mathcal{V}} (u_t(x) - \alpha)^2 \mathrm{d}\mathcal{V} = \frac{1}{2} \|u_t - \alpha\|_{\mathcal{L}_2}^2$ as a Lyapunov functional candidate. Now, it follows from the Green-Gauss theorem and Axiom ii)′ that

$$\begin{aligned}
\dot{\mathcal{E}}_{\mathrm{s}}(u_t) &= \int_{\mathcal{V}} (u(x,t) - \alpha) \frac{\partial u(x,t)}{\partial t} \mathrm{d}\mathcal{V} \\
&= -\int_{\mathcal{V}} u(x,t) \nabla \cdot \phi(x, u(x,t), \nabla u(x,t)) \mathrm{d}\mathcal{V} \\
&\quad + \alpha \int_{\mathcal{V}} \nabla \cdot \phi(x, u(x,t), \nabla u(x,t)) \mathrm{d}\mathcal{V} \\
&= \int_{\mathcal{V}} \nabla u(x,t) \phi(x, u(x,t), \nabla u(x,t)) \mathrm{d}\mathcal{V} \\
&\quad - \int_{\partial \mathcal{V}} u(x,t) \phi(x, u(x,t), \nabla u(x,t)) \cdot \hat{n}(x) \, \mathrm{d}\mathcal{S}_{\mathcal{V}} \\
&\quad + \alpha \int_{\partial \mathcal{V}} \phi(x, u(x,t), \nabla u(x,t)) \cdot \hat{n}(x) \, \mathrm{d}\mathcal{S}_{\mathcal{V}} \\
&= \int_{\mathcal{V}} \nabla u(x,t) \phi(x, u(x,t), \nabla u(x,t)) \mathrm{d}\mathcal{V} \\
&\leq 0, \quad u_t \in \mathcal{W}_2^0(\mathcal{V}), \quad (7.63)
\end{aligned}$$

which establishes Lyapunov stability of the equilibrium state $u(x,t) \equiv \alpha$.

Next, to show semistability of this equilibrium state, consider the following (scaled) ectropy and ectropy-like Lyapunov functionals

$$\mathcal{E}_0(u_t) = \|u_t\|^2_{\mathcal{W}^0_2}, \quad u_t \in \mathcal{W}^0_2(\mathcal{V}), \tag{7.64}$$

$$\mathcal{E}_1(u_t) = \|u_t\|^2_{\mathcal{W}^1_2}, \quad u_t \in \mathcal{W}^1_2(\mathcal{V}). \tag{7.65}$$

It follows from (7.45) with $s(x,t) \equiv 0$ that $\mathcal{E}_0(u_t)$ is a nonincreasing functional of time for all $u_{t_0} \in \mathcal{W}^0_2(\mathcal{V})$. Furthermore, it follows from the Green-Gauss theorem and the boundary condition $\phi(x, u(x,t), \nabla u(x,t)) \cdot \hat{n}(x) \equiv 0$, $x \in \partial\mathcal{V}$, that

$$\begin{aligned}
\tfrac{1}{2}\dot{\mathcal{E}}_1(u_t) = {}& \int_\mathcal{V} \left(u(x,t)\frac{\partial u(x,t)}{\partial t} + \nabla u(x,t)\frac{\partial}{\partial t}\left(\nabla u(x,t)\right)^{\mathrm{T}} \right) \mathrm{d}\mathcal{V} \\
= {}& \int_\mathcal{V} \nabla u(x,t)\phi(x, u(x,t), \nabla u(x,t))\mathrm{d}\mathcal{V} \\
& - \int_{\partial\mathcal{V}} u(x,t)\phi(x, u(x,t), \nabla u(x,t)) \cdot \hat{n}(x)\,\mathrm{d}\mathcal{S}_\mathcal{V} \\
& + \int_{\partial\mathcal{V}} \frac{\partial u(x,t)}{\partial t} D_{\hat{n}(x)}u(x,t)\,\mathrm{d}\mathcal{S}_\mathcal{V} \\
& + \int_\mathcal{V} \nabla^2 u(x,t)\nabla \cdot \phi(x, u(x,t), \nabla u(x,t))\mathrm{d}\mathcal{V} \\
= {}& \int_\mathcal{V} \nabla u(x,t)\phi(x, u(x,t), \nabla u(x,t))\mathrm{d}\mathcal{V} \\
& + \int_{\partial\mathcal{V}} \frac{\partial u(x,t)}{\partial t} D_{\hat{n}(x)}u(x,t)\,\mathrm{d}\mathcal{S}_\mathcal{V} \\
& + \int_\mathcal{V} \nabla^2 u(x,t)\nabla \cdot \phi(x, u(x,t), \nabla u(x,t))\mathrm{d}\mathcal{V}, \tag{7.66}
\end{aligned}$$

where $D_{\hat{n}(x)}u(x,t) \triangleq \nabla u(x,t)\hat{n}(x)$ denotes the directional derivative of $u(x,t)$ along $\hat{n}(x)$ at $x \in \partial\mathcal{V}$. Next, note that for the isolated dynamical system $\mathcal{G}$ with the boundary condition $\phi(x, u(x,t), \nabla u(x,t)) \cdot \hat{n}(x) \equiv 0$, $x \in \partial\mathcal{V}$, it follows from Axiom $i)'$, with $\mathbf{u} = \hat{n}(x)$, that $D_{\hat{n}(x)}u(x,t) \equiv 0$, $x \in \partial\mathcal{V}$. Hence, it follows from Axiom $ii)'$, (7.62), and (7.66) that $\dot{\mathcal{E}}_1(u_t) \leq 0$, $t \geq t_0$, for any $u_{t_0} \in \mathcal{W}^1_2(\mathcal{V})$. Furthermore, since the functionals $\mathcal{E}_1(u_t)$ and $\mathcal{E}_0(u_t)$ are nonincreasing and bounded from below by zero, it follows that $\mathcal{E}_1(u_t)$ and $\mathcal{E}_0(u_t)$ are bounded functionals for every $u_{t_0} \in \mathcal{W}^1_2(\mathcal{V})$. This implies that the positive orbit $\mathcal{O}^+_{u_{t_0}} \triangleq \{u_t \in \mathcal{W}^1_2(\mathcal{V}) : u_t(x) = u(x,t), x \in \mathcal{V}, t \in [t_0, \infty)\}$ of $\mathcal{G}$ is bounded in $\mathcal{W}^1_2(\mathcal{V})$ for all $u_{t_0} \in \mathcal{W}^1_2(\mathcal{V})$. Furthermore, it follows from Sobolev's embedding theorem [92, 98] that $\mathcal{W}^1_2(\mathcal{V})$ is compactly embedded in $\mathcal{W}_2^0(\mathcal{V})$, and hence, $\mathcal{O}_{u_{t_0}}^+$ is contained in a compact subset of $\mathcal{W}_2^0(\mathcal{V})$.

Next, define the sets $\mathcal{D}_{\mathcal{W}_2^1} = \{u_t \in \mathcal{W}_2^1(\mathcal{V}) : \mathcal{E}_1(u_t) < \eta\}$ and $\mathcal{D}_{\mathcal{W}_2^0} = \{u_t \in \mathcal{W}_2^0(\mathcal{V}) : \mathcal{E}_0(u_t) < \eta\}$ for some arbitrary $\eta > 0$. Note that $\mathcal{D}_{\mathcal{W}_2^1}$ and $\mathcal{D}_{\mathcal{W}_2^0}$ are invariant sets with respect to the dynamical system $\mathcal{G}$. Moreover, it follows from the definition of $\mathcal{E}_1(u_t)$ and $\mathcal{E}_0(u_t)$ that $\mathcal{D}_{\mathcal{W}_2^1}$ and $\mathcal{D}_{\mathcal{W}_2^0}$ are bounded sets in $\mathcal{W}_2^1(\mathcal{V})$ and $\mathcal{W}_2^0(\mathcal{V})$, respectively, and $\mathcal{D}_{\mathcal{W}_2^1} \subset \mathcal{D}_{\mathcal{W}_2^0}$. Next, let $\mathcal{R} \triangleq \{u_t \in \overline{\mathcal{D}}_{\mathcal{W}_2^0} : \dot{\mathcal{E}}_0(u_t) = 0\} = \{u_t \in \overline{\mathcal{D}}_{\mathcal{W}_2^0} : \nabla u_t(x)\phi(x, u_t(x), \nabla u_t(x)) = 0, \; x \in \mathcal{V}\}$. Now, it follows from Axioms $i)'$ and $ii)'$ that $\mathcal{R} = \{u_t \in \overline{\mathcal{D}}_{\mathcal{W}_2^0} : \nabla u_t(x) = 0, \; x \in \mathcal{V}\}$ or $\mathcal{R} = \{u_t \in \mathcal{W}_2^0(\mathcal{V}) : u_t(x) \equiv \sigma, \; 0 \le \sigma \le \sqrt{\frac{\eta}{V_{\mathrm{vol}}}}\}$, that is, $\mathcal{R}$ is the set of uniform energy density distributions, which are the equilibrium states of (7.3) and (7.4). Since the set $\mathcal{R}$ consists of only the equilibrium states of (7.3) and (7.4), it follows that the largest invariant set $\mathcal{M}$ contained in $\mathcal{R}$ is given by $\mathcal{M} = \mathcal{R}$. Hence, noting that $\mathcal{M}$ belongs to the set of generalized (or weak) solutions to (7.3) and (7.4) defined on $\mathcal{R}$, it follows from Theorem 2.6 that for any initial energy density distribution $u_{t_0} \in \mathcal{D}_{\mathcal{W}_2^1}$, $u(x, t) \to \mathcal{M}$ as $t \to \infty$ with respect to the norm $\|\cdot\|_{\mathcal{W}_2^0}$, and hence, $u(x, t) \equiv \alpha$ is a semistable equilibrium state of (7.3) and (7.4). Moreover, since $\eta > 0$ can be arbitrarily large but finite and $\mathcal{E}_1(u_t)$ is radially unbounded, the previous statement holds for all $u_{t_0} \in \mathcal{W}_2^1(\mathcal{V})$. Next, note that since, by the divergence theorem,

$$
\begin{aligned}
\int_{\mathcal{V}} \frac{\partial u(x, t)}{\partial t} \mathrm{d}\mathcal{V} &= -\int_{\mathcal{V}} \nabla \cdot \phi(x, u(x, t), \nabla u(x, t)) \mathrm{d}\mathcal{V} \\
&= -\int_{\partial \mathcal{V}} \phi(x, u(x, t), \nabla u(x, t)) \cdot \hat{n}(x) \, \mathrm{d}\mathcal{S}_{\mathcal{V}} \\
&= 0,
\end{aligned}
\tag{7.67}
$$

it follows that $\int_{\mathcal{V}} u(x, t) \mathrm{d}\mathcal{V} = \int_{\mathcal{V}} u_{t_0}(x) \mathrm{d}\mathcal{V}$, $t \ge t_0$, which implies that $u(x, t) \to \frac{1}{V_{\mathrm{vol}}} \int_{\mathcal{V}} u_{t_0}(x) \mathrm{d}\mathcal{V}$ as $t \to \infty$.

Finally, we show that if $s(x, t) \equiv 0$ and there exists at least one point $x_{\mathrm{p}} \in \partial \mathcal{V}$ such that $\phi(x_{\mathrm{p}}, u_t(x_{\mathrm{p}}), \nabla u_t(x_{\mathrm{p}})) \cdot \hat{n}(x_{\mathrm{p}}) > 0$ and $\phi(x_{\mathrm{p}}, u_t(x_{\mathrm{p}}), \nabla u_t(x_{\mathrm{p}})) \cdot \hat{n}(x_{\mathrm{p}}) = 0$ if and only if $u_t(x_{\mathrm{p}}) = 0$, then the zero solution $u(x, t) \equiv 0$ to (7.3) and (7.4) is a globally asymptotically stable equilibrium state. Note that it follows from the above analysis with $\alpha = 0$ that the zero solution $u(x, t) \equiv 0$ is semistable, and hence, a Lyapunov stable equilibrium state of (7.3) and (7.4). Furthermore, it follows from Axiom $ii)'$ with $\mathbf{u} = \hat{n}(x_{\mathrm{p}})$ that

$D_{\hat{n}(x_{\mathrm{p}})}u(x_{\mathrm{p}}, t) = \nabla u(x_{\mathrm{p}}, t)\hat{n}(x_{\mathrm{p}}) < 0$ and $D_{\hat{n}(x_{\mathrm{p}})}u(x_{\mathrm{p}}, t) = 0$ if and only if $u(x_{\mathrm{p}}, t) = 0$. In this case, using Axiom $ii)'$, it follows that the energy flow is directed towards the point $x_{\mathrm{p}} \in \partial\mathcal{V}$, and hence, $\frac{\partial u(x_{\mathrm{p}}, t)}{\partial t} > 0$ and $D_{\hat{n}(x_{\mathrm{p}})}u(x_{\mathrm{p}}, t)\frac{\partial u(x_{\mathrm{p}}, t)}{\partial t} < 0$. Thus, it follows from Axiom $ii)'$, (7.62), and (7.66) that $\mathcal{E}_1(u_t)$ is a nonincreasing functional of time for all $u_{t_0} \in \mathcal{W}_2^1(\mathcal{V})$, and since $\mathcal{E}_1(u_t)$ is bounded from below by zero, the positive orbit $\mathcal{O}_{u_{t_0}}^+$ of $\mathcal{G}$ is bounded in $\mathcal{W}_2^1(\mathcal{V})$. Hence, since $\mathcal{W}_2^1(\mathcal{V})$ is compactly embedded in $\mathcal{W}_2^0(\mathcal{V})$, it follows from Sobolev's embedding theorem [92, 98] that $\mathcal{O}_{u_{t_0}}^+$ is contained in a compact subset of $\mathcal{W}_2^0(\mathcal{V})$.

Next, consider the (scaled) ectropy Lyapunov functional $\mathcal{E}_0(u_t)$ and note that the Lyapunov derivative is given by

$$
\begin{aligned}
\tfrac{1}{2}\dot{\mathcal{E}}_0(u_t) &= \int_{\mathcal{V}} u(x, t)\frac{\partial u(x, t)}{\partial t}\mathrm{d}\mathcal{V} \\
&= -\int_{\mathcal{V}} u(x, t)\nabla \cdot \phi(x, u(x, t), \nabla u(x, t))\mathrm{d}\mathcal{V} \\
&= \int_{\mathcal{V}} \nabla u(x, t)\phi(x, u(x, t), \nabla u(x, t))\mathrm{d}\mathcal{V} \\
&\quad - \int_{\partial\mathcal{V}} u(x, t)\phi(x, u(x, t), \nabla u(x, t)) \cdot \hat{n}(x)\,\mathrm{d}\mathcal{S}_{\mathcal{V}} \\
&\leq 0, \quad u_t \in \mathcal{W}_2^0(\mathcal{V}).
\end{aligned}
\tag{7.68}
$$

Furthermore, let $\mathcal{R} \triangleq \{u_t \in \overline{\mathcal{D}}_{\mathcal{W}_2^0} : \dot{\mathcal{E}}_0(u_t) = 0\} = \{u_t \in \overline{\mathcal{D}}_{\mathcal{W}_2^0} : \nabla u_t(x)\phi(x, u_t(x), \nabla u_t(x)) \equiv 0, \ x \in \mathcal{V}\} \cap \{u_t \in \overline{\mathcal{D}}_{\mathcal{W}_2^0} : \phi(x, u_t(x), \nabla u_t(x)) \cdot \hat{n}(x) = 0, \ x \in \partial\mathcal{V}\}$. Now, since Axioms $i)'$ and $ii)'$ hold, $\mathcal{R} = \{u_t \in \overline{\mathcal{D}}_{\mathcal{W}_2^0} : \nabla u_t(x) = 0, \ x \in \mathcal{V}\} \cap \{u_t \in \overline{\mathcal{D}}_{\mathcal{W}_2^0} : u_t(x_{\mathrm{p}}) = 0 \text{ for some } x_{\mathrm{p}} \in \partial\mathcal{V}\} = \{0\}$, and the largest invariant set $\mathcal{M}$ contained in $\mathcal{R}$ is given by $\mathcal{M} = \{0\}$. Hence, it follows from Theorem 2.6 that for any initial energy density distribution $u_{t_0} \in \mathcal{D}_{\mathcal{W}_2^1}$, $u(x, t) \to \mathcal{M} = \{0\}$ as $t \to \infty$ with respect to the norm $\|\cdot\|_{\mathcal{W}_2^0}$, which, since $\eta > 0$ is arbitrary and $\mathcal{E}_1(u_t)$ is radially unbounded, proves global asymptotic stability of the zero equilibrium state of (7.3) and (7.4). $\quad\square$

Condition (7.62) physically implies that for an energy density distribution $u_t(x)$, $x \in \mathcal{V}$, the energy flow $\phi(x, u_t(x), \nabla u_t(x))$ at $x \in \mathcal{V}$ is proportional to the energy density at this point. Note that for the linear energy flow model corresponding to the heat equation, that is, $\phi(x, u_t(x), \nabla u_t(x)) = -k[\nabla u_t(x)]^{\mathrm{T}}$, where $k > 0$ is a conductivity constant, condition (7.62) is automatically satisfied since

$\nabla^2 u_t(x) \nabla \cdot \phi(x, u_t(x), \nabla u_t(x)) = -k[\nabla^2 u_t(x)]^2 \leq 0$, $x \in \mathcal{V}$. Theorem 7.9 shows that the isolated dynamical system $\mathcal{G}$ is semistable. Hence, it follows from the infinite-dimensional version of Theorem 2.16 that the isolated dynamical system $\mathcal{G}$ does not exhibit Poincaré recurrence in $\mathcal{X} \setminus \mathcal{M}_e$. This result can also be arrived at using (7.39) or (7.53) along with the infinite-dimensional version of Theorem 2.15.

Next, we give an analogous proposition to Proposition 3.10 for infinite-dimensional systems.

Proposition 7.7 *Consider the dynamical system $\mathcal{G}$ with power balance equation (7.3) and (7.4), let $\mathcal{E} : \mathcal{X} \to \overline{\mathbb{R}}_+$ and $\mathcal{S} : \mathcal{X} \to \overline{\mathbb{R}}_+$ denote the ectropy and entropy functionals of $\mathcal{G}$ given by (7.52) and (7.38), respectively, and define $\mathcal{D}_c \triangleq \{u_t \in \mathcal{X} : \int_{\mathcal{V}} u_t(x) d\mathcal{V} = \beta\}$, where $\beta \geq 0$. Then*

$$\arg\min_{u_t \in \mathcal{D}_c}(\mathcal{E}(u_t)) = \arg\max_{u_t \in \mathcal{D}_c}(\mathcal{S}(u_t)) = u_t^* = \frac{\beta}{\mathcal{V}_{\text{vol}}}. \qquad (7.69)$$

Furthermore, $\mathcal{E}_{\min} \triangleq \mathcal{E}(u_t^*) = \frac{\beta^2}{2}$ and $\mathcal{S}_{\max} \triangleq \mathcal{S}(u_t^*) = \mathcal{V}_{\text{vol}}[\log_e(c + \frac{\beta}{\mathcal{V}_{\text{vol}}}) - \log_e c]$.

Proof. The proof is similar to the proof of Proposition 3.10 and, hence, is omitted. The only difference here is that $\mathcal{E}(u_t)$ and $-\mathcal{S}(u_t)$ are real-valued convex *functionals* defined on $\mathcal{X}$, and $\int_{\mathcal{V}} u_t(x) d\mathcal{V}$ is a convex mapping from a convex subset of $\mathcal{X}$ into a normed space. The result thus follows as a direct consequence of global theory for constrained optimization of functionals [68]. $\qquad\square$

Next, we use the entropy functional (respectively, ectropy functional) given by (7.38) (respectively, (7.52)) to show a clear connection between our continuum thermodynamic model given by (7.3) and (7.4), and the arrow of time.

Theorem 7.10 *Consider the dynamical system $\mathcal{G}$ with power balance equation (7.3) and (7.4) with $s(x, t) \equiv 0$ and $\phi(x, u(x, t), \nabla u(x, t)) \cdot \hat{n}(x) \equiv 0$, $x \in \partial\mathcal{V}$, and assume Axioms i)' and ii)' hold. Furthermore, let $s^u(\cdot, 0) \in \mathcal{W}_{[t_0, t_1]}$. Then for every $u_{t_0} \notin \mathcal{M}_e$, there exists a continuously differentiable functional $\mathcal{S} : \mathcal{X} \to \mathbb{R}$ (respectively, $\mathcal{E} : \mathcal{X} \to \mathbb{R}$) such that $\mathcal{S}(s^u(t, 0))$ (respectively, $\mathcal{E}(s^u(t, 0))$) is a strictly increasing (respectively, decreasing) function of time. Furthermore, $s^u(\cdot, 0)$ is an $I_{\mathcal{X}}$-reversible trajectory of $\mathcal{G}$ if and only if $s^u(t, 0) \in \mathcal{M}_e$, $t \in [t_0, t_1]$.*

Proof. The proof is similar to the proof of Theorem 3.10 and follows from Corollary 2.4. $\qquad\square$

We close this section by noting that the results of this chapter can be easily generalized to the case where the energy density at a point $x \in \mathcal{V}$ is proportional to the temperature, that is, $\hat{T}(x, t) = \beta(x)u(x, t)$, where $\hat{T}(x, t)$ is the temperature distribution over the continuum and $\beta(x)$ is the reciprocal of the specific heat (thermal capacity) at the spatial coordinate x. In this case, analogous results to the results of Section 4.1 can be easily derived for the infinite-dimensional thermodynamic model. Finally, it is important to note that the results of this section apply to an arbitrary (not necessarily Cartesian) n-dimensional space. In particular, we could consider a coordinate transformation $y = Y(x)$, where $Y(0) = 0$ and $Y : \mathcal{V} \to \mathbb{R}^n$ is a diffeomorphism in the neighborhood of the origin, so that y is defined on the image of $\mathcal{V} \subset \mathbb{R}^n$ under the mapping Y. In this case, however, the nabla and gradient operators need to be redefined appropriately [2, pp. 350–351].

Chapter Eight

Conclusion

In this monograph, we have outlined a general systems theory framework for thermodynamics in an attempt to harmonize it with classical mechanics. The proposed macroscopic mathematical model is based on a nonlinear (finite- and infinite-dimensional) compartmental dynamical system model that is characterized by energy conservation laws capturing the exchange of energy between coupled macroscopic subsystems. Specifically, using a large-scale dynamical systems perspective, we developed some of the fundamental properties of reversible and irreversible thermodynamic systems involving conservation of energy, nonconservation of entropy and ectropy, and energy equipartition. This model is formulated in the language of dynamical systems and control theory, and it is argued that it offers conceptual advantages for describing nonequilibrium thermodynamic systems.

Using compartmental dynamical systems involving the exchange of energy via intercompartmental flow laws and invoking the two fundamental axioms of the science of heat, namely,

> i) *if the energies in the connected subsystems are equal, then energy exchange between these subsystems is not possible,*

and

> ii) *energy flows from more energetic subsystems to less energetic subsystems,*

we established the existence of a continuous entropy function for our thermodynamically consistent large-scale dynamical system utilizing the language of modern mathematics within a theorem-proof format. In addition, we prove the global existence and uniqueness of a continuously differentiable entropy and ectropy function for all equilibrium and nonequilibrium states of our dynamical system. Furthermore, the fundamental properties of reversible and irreversible thermodynamics were also established using a system-theoretic dynamical systems approach.

In particular, for our thermodynamically consistent large-scale dynamical system, it was shown that:

   i) *The increase in internal energy of a dynamical system equals the heat energy received by the system minus the work expended by the system.*

  ii) *The total energy in an isolated dynamical system is constant.*

 iii) *For every dynamical transformation in an adiabatically isolated system, the entropy of the final state is greater than or equal to the entropy of the initial state.*

  iv) *The entropy of an adiabatically isolated dynamical system tends to a maximum.*

   v) *An isolated large-scale dynamical system naturally evolves toward a state of energy equipartition.*

  vi) *Although the total energy in an adiabatically isolated dynamical system is conserved, the usable energy is diffused.*

 vii) *For an equilibrium of any isolated dynamical system, it is necessary and sufficient that in all possible variations of the state of the system that do not alter its energy, the change in entropy is zero or negative.*

viii) *The entropy of every dynamical system at absolute zero can always be taken to be equal to zero.*

In addition, in our formulation the notion of subsystem thermodynamic temperatures is derived as a direct consequence of the existence of the unique, continuously differentiable subsystem entropies. Hence, thermal equilibrium is an equivalence relation between subsystem energies and does not rely on the subjective notions of hotness and coldness of each subsystem.

   In this monograph, we have largely concentrated on classical thermodynamics with little mention of statistical mechanics. However, as noted in Chapter 1, the theory of thermodynamics followed two conceptually rather different schools of thought, namely, the macroscopic point of view versus the microscopic point of view. The microscopic point of view of thermodynamics was first established by Maxwell [72] and further developed by Boltzmann [15] by reinterpreting thermodynamic systems in terms of molecules or atoms. However, since the microscopic states of thermodynamic systems involve a large number of similar molecules, the laws of classical mechanics were reformulated so that even though individual atoms are assumed to obey the laws of Newtonian mechanics, the statistical nature of the velocity distribution of the system particles corresponds to the thermodynamic properties of all the atoms together. This resulted in the birth of statistical mechanics. The laws of mechanics, however, as established by Poincaré [32], show that every isolated mechanical system will return arbitrarily close to its initial state infinitely often. Hence, entropy must undergo cyclic changes and thus cannot increase. This is known as the *recurrence paradox* or *Loschmidt's paradox*.

Loschmidt [66] was among the first to challenge the theory of statistical thermodynamics by pointing out that Boltzmann's theory violated the time-reversal symmetry of the microscopic equations of motion of the system particles. In fact, Poincaré's recurrence theorem prohibits irreversibility of conservative dynamical systems in the classical sense. To the present day, many scientists have attempted to provide an explanation of the recurrence paradox in which a lossless dynamical system that possesses time-reversal symmetry on a microscopic scale breaks this symmetry on a macroscopic scale. Many scientists have made untenable arguments that despite microscopic reversibility, not all solutions need possess full time-reversal symmetry while others have averted their eyes from Loschmidt's paradox. In light of Poincaré recurrence, the law of entropy increase cannot be derived from statistical mechanics and to this point has eluded the deepest thinkers in science. The problem of duplicating the second law of thermodynamics remains one of the hardest and most controversial problems in statistical physics.

In statistical thermodynamics the recurrence paradox is resolved by asserting that, in principle, the entropy of an isolated system can sometimes decrease. However, the probability of this happening, when computed, is incredibly small. Thus, statistical thermodynamics stipulates that the direction in which system transformations occur is determined by the laws of probability, and hence, they result in a more probable state corresponding to a higher system entropy. However, unlike classical thermodynamics, in statistical thermodynamics it is not absolutely certain that entropy increases in every system transformation. Hence, thermodynamics based on statistical mechanics gives the most probable course of system evolution and not the only possible one, and thus heat flows in the direction of lower temperature with only *statistical certainty* and not absolute certainty. Nevertheless, general arguments exploiting system fluctuations in a systematic way [99] seem to show that it is impossible, even in principle, to violate the second law of thermodynamics. In fact, no exception has ever been found to the second law of thermodynamics making it, along with the first law, one of the most perfect laws of nature.

In this regard, Eddington [37, p. 81] writes:

> The law that entropy always increases—the second law of thermo-dynamics—holds, I think, the supreme position among the laws of Nature. If someone points out to you that your pet theory of the universe is in disagreement with Maxwell's equations—then so much worse for Maxwell's equations. If it is found to be contradicted by observation—well, these experimentalists bungle things sometimes. But if your theory is found to be against the second law of thermodynamics I can give you no hope; there is nothing for it but to collapse in deepest humiliation.

The underlying intention of this monograph has been to present one of the most useful and general physical branches of science in the language of dynamical systems theory. In particular, we developed a novel formulation of thermodynamics using a middle-ground systems theory that bridges the gap between classical and statistical thermodynamics. The laws of thermodynamics are among the most firmly established laws of nature, and it is hoped that this monograph will help to stimulate increased interaction between physicists and dynamical systems and control theorists. Besides the fact that irreversible thermodynamics plays a critical role in the understanding of our expanding universe, it forms the underpinning of several fundamental life science and engineering disciplines, including biological systems, physiological systems, chemical reaction systems, queuing systems, ecological systems, demographic systems, telecommunications systems, transportation systems, network systems, and power systems, to cite but a few examples.

The newly developed dynamical system notion of entropy proposed in this monograph involving an analytical description of an objective property of matter can potentially offer a conceptual advantage over the subjective quantum expressions for entropy proposed in the literature (e.g., Daróczy entropy, Hartley entropy, Rényi entropy, von Neumann entropy, infinite-norm entropy) involving a measure of information. An even more important benefit of the dynamical system representation of thermodynamics is the potential for developing a unified classical and quantum theory that encompasses both mechanics and thermodynamics without the need for statistical (subjective or informational) probabilities.

There is no doubt that thermodynamics is a theory of universal proportions whose laws reign supreme among the laws of nature and are capable of addressing some of science's most intriguing questions about the origins and fabric of our universe. While from its inception its speculations about the universe have been grandiose, its mathematical foundation has been amazingly obscure and imprecise. A discipline as cardinal as thermodynamics entrusted with some of the most perplexing secrets of our universe demands far more than "physical" mathematics as its underpinning. Even though many great physicists such as Archimedes, Newton, and Lagrange have humbled us with their mathematically seamless eurekas over the centuries, a great many physicists and engineers who have developed the theory of thermodynamics over the last one and a half centuries seem to have forgotten that mathematics, when used rigorously, is the irrefutable pathway to truth.

Our goal with this monograph has been to develop a dynamical system formalism for classical thermodynamics. As a result, we use system theoretic ideas to bring coherence, clarity, and precision to an extremely important and poorly understood classical area of science. Our systems thermodynamics formalism brings classical thermodynamics within the framework of modern dynamical systems by bringing to bear some of the hallmark analytical tools from dynamical systems and control theory. A dynamical system formalism of thermodynamics has been long overdue and aligns classical thermodynamics with the development of classical mechanics, which also started as a physical theory concerned mainly with equilibrium systems and with empirical principles initially formulated by the great cosmic theorists of ancient Greece, and later established by physicists such as Copernicus, Brahe, Kepler, and Galileo. However, unlike classical thermodynamics which remained a physical theory, in the seventeenth through the nineteenth centuries the physical approach of mechanics was replaced by mathematical theories involving abstract geometrical structures (configuration manifolds, Riemann space, Minkowski space-time), wherein the mechanistic empirical principles were incorporated into topological properties of abstract mathematical spaces. This physical-mathematical bifurcation of mechanics, which was pioneered by giants such as Newton, Huygens, Lagrange, and Hamilton, along with the fact that classical thermodynamics remained concerned with systems in equilibrium, made it all but impossible to unify classical thermodynamics with classical mechanics, leaving these two classical disciplines of physics to stand in sharp contrast to one another in the one and a half centuries of their coexistence.

While it seems impossible to reduce thermodynamics to a mechanistic world picture due to microscopic reversibility and Poincaré recurrence, our system thermodynamic formulation provides a harmonization of classical thermodynamics with classical mechanics. In particular, our dynamical system formalism captures all of the key aspects of thermodynamics, including its fundamental laws, while providing a mathematically rigorous formulation for thermodynamical systems out of equilibrium by unifying the theory of heat transfer with that of classical thermodynamics. In addition, the concept of entropy for a nonequilibrium state of a dynamical process is defined, and its global existence and uniqueness is established. This state space formalism of thermodynamics shows that the behavior of heat, as described by the conservation equations of thermal transport and as described by classical thermodynamics, can be derived from the same basic principles and is part of the same scientific discipline. Finally, classical thermodynamics meets Fourier's theory of heat conduction in the one hundred and fifty years of their coexistence. And for those numerous thermodynamicists who have repeatedly confused statics with dynamics and have been under the illusion that their science of *thermostatics* (i.e., classical thermodynamics) somehow rivals classical mechanics, in consequence, they need not so remain.

Bibliography

[1] R. Abraham, J. E. Marsden, and T. Ratin, *Manifolds, Tensor Analysis, and Applications.* New York, NY: Springer-Verlag, 1988.

[2] T. M. Apostol, *Mathematical Analysis.* Reading, MA: Addison-Wesley, 1957.

[3] T. M. Apostol, *Mathematical Analysis*, 2nd ed. Reading, MA: Addison-Wesley, 1974.

[4] V. I. Arnold, *Mathematical Models of Classical Mechanics.* New York, NY: Springer-Verlag, 1989.

[5] V. I. Arnold, "Contact geometry: The geometrical method of Gibbs' thermodynamics," in *Proceedings of the Gibbs Symposium* (D. Caldi and G. Mostow, eds.), pp. 163–179. Providence, RI: American Mathematical Society, 1990.

[6] A. Berman and R. J. Plemmons, *Nonnegative Matrices in the Mathematical Sciences.* New York, NY: Academic Press, Inc., 1979.

[7] A. Berman, R. S. Varga, and R. C. Ward, "ALPS: Matrices with nonpositive off-diagonal entries," *Linear Algebra Appl.*, vol. 21, pp. 233–244, 1978.

[8] D. S. Bernstein and S. P. Bhat, "Nonnegativity, reducibilty and semistability of mass action kinetics," in *Proc. IEEE Conf. Dec. Contr.* (Phoenix, AZ), pp. 2206–2211, 1999.

[9] D. S. Bernstein and S. P. Bhat, "Energy equipartition and the emergence of damping in lossless systems," in *Proc. IEEE Conf. Dec. Contr.* (Las Vegas, NV), pp. 2913–2918, 2002.

[10] D. S. Bernstein and D. C. Hyland, "Compartmental modeling and second-moment analysis of state space systems," *SIAM J. Matrix Anal. Appl.*, vol. 14, pp. 880–901, 1993.

[11] J. Bertrand, *Thermodynamique*. Paris: Gauthier-Villars, 1887.

[12] S. P. Bhat and D. S. Bernstein, "Lyapunov analysis of semistability," in *Amer. Contr. Conf.* (San Diego, CA), pp. 1608–1612, 1999.

[13] S. P. Bhat and D. S. Bernstein, "Nontangency-based Lyapunov tests for convergence and stability in systems having a continuum of equilibria," *SIAM J. Control Optim.*, vol. 42, pp. 1745–1775, 2003.

[14] N. P. Bhatia and G. P. Szegö, *Stability Theory of Dynamical Systems*. Berlin: Springer-Verlag, 1970.

[15] L. Boltzmann, *Vorlesungen über die Gastheorie*, 2nd ed. Leipzig: J. A. Barth, 1910.

[16] P. Bridgman, *The Nature of Thermodynamics*. Cambridge, MA: Harvard University Press, 1941. Reprinted (Gloucester, MA: Peter Smith, 1969).

[17] R. W. Brockett and J. C. Willems, "Stochastic control and the second law of thermodynamics," in *Proc. IEEE Conf. Dec. Contr.* (San Diego, CA), pp. 1007–1011, 1978.

[18] J. Brunet, "Information theory and thermodynamics," *Cybernetica*, vol. 32, pp. 45–78, 1989.

[19] S. G. Brush, *The Kind of Motion we Call Heat: A History of the Kinetic Theory in the Nineteenth Century*. Amsterdam: North Holland, 1976.

[20] C. Carathéodory, "Untersuchungen über die Grundlagen der Thermodynamik," *Math. Annalen*, vol. 67, pp. 355–386, 1909.

[21] C. Carathéodory, "Über die Bestimmung der Energie und der absoluten Temperatur mit Hilfe von reversiblen Prozessen," *Sitzungsberichte der preußischen Akademie der Wissenschaften, Math. Phys. Klasse*, pp. 39–47, 1925.

[22] A. Carcaterra, "An entropy formulation for the analysis of energy flow between mechanical resonators," *Mechanical Systems and Signal Processing*, vol. 16, pp. 905–920, 2002.

[23] D. S. L. Cardwell, *From Watt to Clausius: The Rise of Thermodynamics in the Early Industrial Age*. Ithaca, NY: Cornell University Press, 1971.

[24] S. Carnot, *Réflexions sur la puissance motrice du feu et sur les machines propres a développer cette puissance.* Paris: Chez Bachelier, Libraire, 1824.

[25] H. B. G. Casimir, "On Onsager's principle of microscopic reversibility," *Rev. Mod. Phys.*, vol. 17, pp. 343–350, 1945.

[26] R. Clausius, "Über die Concentration von Wärme- und Lichtstrahlen und die Gränze Ihre Wirkung," in *Abhandlungen über die Mechanischen Wärmetheorie*, pp. 322–361. Braunschweig: Vieweg & Sohn, 1864.

[27] R. Clausius, "Über verschiedene für die Anwendung bequeme Formen der Haubtgleichungen der mechanischen wärmetheorie," *Vierteljahrschrift der naturforschenden Gesellschaft (Zürich)*, vol. 10, pp. 1–59, 1865, also in [29, pp. 1–56], and translated in [55, pp. 162–193].

[28] R. Clausius, *Abhandlungungen über die Mechanische Wärmetheorie*, vol. 2. Braunschweig: Vieweg & Sohn, 1867.

[29] R. Clausius, *Die Mechanische Wärmetheorie.* Braunschweig: Vieweg & Sohn, 1876.

[30] B. D. Coleman, "The thermodynamics of materials with memory," *Arch. Rational Mech. Anal.*, vol. 17, pp. 1–46, 1964.

[31] B. D. Coleman and W. Noll, "The thermodynamics of elastic materials with heat conduction and viscosity," *Arch. Rational Mech. Anal.*, vol. 13, pp. 167–178, 1963.

[32] P. Coveney, *The Arrow of Time.* New York, NY: Ballantine Books, 1990.

[33] C. M. Dafermos, *Hyperbolic Conservation Laws in Continuum Physics.* Berlin: Springer-Verlag, 2000.

[34] W. A. Day, "Thermodynamics based on a work axiom," *Arch. Rational Mech. Anal.*, vol. 31, pp. 1–34, 1968.

[35] W. A. Day, "A theory of thermodynamics for materials with memory," *Arch. Rational Mech. Anal.*, vol. 34, pp. 86–96, 1969.

[36] J. Earman, "Irreversibility and temporal asymmetry," *J. Philos.*, vol. 64, pp. 543–549, 1967.

[37] A. Eddington, *The Nature of the Physical World.* London: J. M. Dent & Sons, 1935.

[38] L. C. Evans, *Partial Differential Equations.* Providence, RI: American Mathematical Society, 1998.

[39] J. W. Gibbs, *The Scientific Papers of J. Willard Gibbs:* Vol. 1, *Thermodynamics.* London: Longmans, 1906.

[40] R. Giles, *Mathematical Foundations of Thermodynamics.* Oxford: Pergamon, 1964.

[41] M. Goldstein and I. F. Goldstein, *The Refrigerator and the Universe.* Cambridge, MA: Harvard University Press, 1993.

[42] A. Greven, G. Keller, and G. Warnecke, eds., *Entropy.* Princeton, NJ: Princeton University Press, 2003.

[43] A. Grünbaum, "The anisotropy of time," in *The Nature of Time* (T. Gold, ed.), pp. 149–186. Ithaca, NY: Cornell University Press, 1967.

[44] M. Gurtin, "On the thermodynamics of materials with memory," *Arch. Rational Mech. Anal.*, vol. 28, pp. 40–50, 1968.

[45] E. P. Gyftopoulos and G. P. Beretta, *Thermodynamics: Foundations and Applications.* New York, NY: Macmillan, 1991.

[46] E. P. Gyftopoulos and E. Çubukçu, "Entropy: Thermodynamic definition and quantum expression," *Phys. Rev. E*, vol. 55, no. 4, pp. 3851–3858, 1997.

[47] W. M. Haddad and V. Chellaboina, "Stability and dissipativity theory for nonnegative dynamical systems: A unified analysis framework for biological and physiological systems," *Nonlinear Analysis: Real World Applications*, vol. 6, pp. 35–65, 2005.

[48] W. M. Haddad, V. Chellaboina, and E. August, "Stability and dissipativity theory for nonnegative dynamical systems: A thermodynamic framework for biological and physiological systems," in *Proc. IEEE Conf. Dec. Contr.* (Orlando, FL), pp. 442–458, 2001.

[49] J. K. Hale, "Dynamical systems and stability," *J. Math. Anal. Appl.*, vol. 26, pp. 39–59, 1969.

[50] J. K. Hale, *Ordinary Differential Equations*, 2nd ed. New York: Wiley, 1980. Reprinted (Malabar: Krieger, 1991).

[51] S. R. Hall, D. G. MacMartin, and D. S. Bernstein, "Covariance averaging in the analysis of uncertain systems," in *Proc. IEEE Conf. Dec. Contr.* (Tucson, AZ), pp. 1842–1859, 1992.

[52] D. J. Hill and P. J. Moylan, "Dissipative dynamical systems: Basic input-output and state properties," *J. Franklin Inst.*, vol. 309, pp. 327–357, 1980.

[53] P. Horwich, *Asymmetries in Time*. Cambridge, MA: MIT Press, 1987.

[54] A. J. Keane and W. G. Price, "Statistical energy analysis of strongly coupled systems," *J. Sound Vibration*, vol. 117, pp. 363–386, 1987.

[55] J. Kestin, *The Second Law of Thermodynamics*. Stroudsburg, PA: Dowden, Hutchinson and Ross, 1976.

[56] H. K. Khalil, *Nonlinear Systems*. Upper Saddle River, NJ: Prentice-Hall, 1996.

[57] Y. Kishimoto and D. S. Bernstein, "Thermodynamic modeling of interconnected systems, I: Conservative coupling," *J. Sound Vibration*, vol. 182, pp. 23–58, 1995.

[58] Y. Kishimoto and D. S. Bernstein, "Thermodynamic modeling of interconnected systems, II: Dissipative coupling," *J. Sound Vibration*, vol. 182, pp. 59–76, 1995.

[59] Y. Kishimoto, D. S. Bernstein, and S. R. Hall, "Energy flow modeling of interconnected structures: A deterministic foundation for statistical energy analysis," *J. Sound Vibration*, vol. 186, pp. 407–445, 1995.

[60] P. Kroes, *Time: Its Structure and Role in Physical Theories*. Dordrecht: D. Reidel, 1985.

[61] J. S. W. Lamb and J. A. G. Roberts, "Time reversal symmetry in dynamical systems: A survey," *Phys. D*, vol. 112, pp. 1–39, 1998.

[62] R. S. Langley, "A general derivation of the statistical energy analysis equations for coupled dynamic systems," *J. Sound Vibration*, vol. 135, pp. 499–508, 1989.

[63] B. Lavenda, *Thermodynamics of Irreversible Processes*. London: Macmillan, 1978. Reprinted (New York: Dover, 1993).

[64] P. Liberman and C. M. Marle, *Symplectic Geometry and Analytical Mechanics*. Dordrecht: Reidel, 1987.

[65] E. H. Lieb and J. Yngvason, "The physics and mathematics of the second law of thermodynamics," *Phys. Rep.*, vol. 310, pp. 1–96, erratum, vol. 314, p. 669, 1999.

[66] J. Loschmidt, "Über den Zustand des Wärmegleichgewichtes eines Systems von Körpern mit Rücksicht auf die Schwere," *Wiener Berichte*, vol. 73, pp. 128–142, 1876.

[67] R. Lozano, B. Brogliato, O. Egeland, and B. Maschke, *Dissipative Systems Analysis and Control*. London: Springer-Verlag, 2000.

[68] D. G. Luenberger, *Optimization by Vector Space Methods*. New York, NY: Wiley, 1969.

[69] R. H. Lyon, *Statistical Energy Analysis of Dynamical Systems: Theory and Applications*. Cambridge, MA: MIT Press, 1975.

[70] M. C. Mackey, *Time's Arrow: The Origins of Thermodynamic Behavior*. New York, NY: Springer-Verlag, 1992.

[71] M. Marvan, *Negative Absolute Temperatures*. London: Iliffe Books, 1966.

[72] J. C. Maxwell, "On the dynamical theory of gases," *Philos. Trans. Roy. Soc. London Ser. A*, vol. 157, pp. 49–88, 1866.

[73] J. Meixner, "On the foundation of thermodynamics of processes," in *A Critical Review of Thermodynamics* (E. B. Stuart, B. Gal-Or, and A. J. Brainard, eds.), pp. 37–47. Baltimore, MD: Mono Book Corp., 1970.

[74] N. Minamide, "An extension of the matrix inversion lemma," *SIAM J. Alg. Disc. Meth.*, vol. 6, pp. 371–377, 1985.

[75] I. Müller, "Die Kältefunktion, eine universelle Funktion in der Thermodynamik viscoser wärmeleitender Flüssigkeiten," *Arch. Rational Mech. Anal.*, vol. 40, pp. 1–36, 1971.

[76] H. Nijmeijer and A. J. van der Schaft, *Nonlinear Dynamical Control Systems*. New York, NY: Springer, 1990.

[77] E. F. Obert, *Concepts of Thermodynamics*. New York, NY: McGraw-Hill, 1960.

[78] L. Onsager, "Reciprocal relations in irreversible processes, I," *Phys. Rev.*, vol. 37, pp. 405–426, 1931.

[79] L. Onsager, "Reciprocal relations in irreversible processes, II," *Phys. Rev.*, vol. 38, pp. 2265–2279, 1932.

[80] M. Pavon, "Stochastic control and nonequilibrium thermodynamical systems," *Appl. Math. Optim.*, vol. 19, pp. 187–202, 1989.

[81] R. K. Pearson and T. L. Johnson, "Energy equipartition and fluctuation-dissipation theorems for damped flexible structures," *Quart. Appl. Math.*, vol. 45, pp. 223–238, 1987.

[82] M. Planck, *Vorlesungen über Thermodynamik*. Leipzig: Veit, 1897.

[83] M. Planck, "Über die Begrundung des zweiten Hauptsatzes der Thermodynamik," *Sitzungsberichte der preußischen Akademie der Wissenschaften, Math. Phys. Klasse*, pp. 453–463, 1926.

[84] H. Poincaré, "Sur le probléme des trois corps et les équations de la dynamique," *Acta Math.*, vol. 13, pp. 1–270, 1890.

[85] I. Prigogine, *From Being to Becoming*. San Francisco: W. H. Freeman, 1980.

[86] N. Ramsey, "Thermodynamics and statistical mechanics at negative absolute temperatures," *Phys. Rev.*, vol. 103, p. 20, 1956.

[87] H. Reichenbach, *The Direction of Time*. Berkeley, CA: University of California Press, 1956.

[88] H. L. Royden, *Real Analysis*. Englewood Cliffs, NJ: Prentice-Hall, 1988.

[89] R. G. Sachs, *The Physics of Time Reversal*. Chicago, IL: University of Chicago Press, 1987.

[90] H. L. Smith, *Monotone Dynamical Systems*. Providence, RI: American Mathematical Society, 1995.

[91] P. W. Smith, "Statistical models of coupled dynamical systems and the transition from weak to strong coupling," *Journal of the Acoustical Society of America*, vol. 65, pp. 695–698, 1979.

[92] S. L. Sobolev, "Applications of functional analysis in mathematical physics," *Translations of Mathematical Monographs*, vol. 7. Providence, RI: American Mathematical Society, 1963.

[93] W. Thomson (Lord Kelvin), "On a universal tendency in nature to the dissipation of mechanical energy," *Proc. Roy. Soc. Edinburgh*, vol. 20, pp. 139–142, 1852.

[94] L. Tisza, "Thermodynamics in a state of flux. A search for new foundations," in *A Critical Review of Thermodynamics* (E. B. Stuart, B. Gal-Or, and A. J. Brainard, eds.), pp. 107–118. Baltimore, MD: Mono Book Corp., 1970.

[95] C. Truesdell, *Rational Thermodynamics.* New York, NY: McGraw-Hill, 1969.

[96] C. Truesdell, *The Tragicomical History of Thermodynamics 1822–1854.* New York, NY: Springer-Verlag, 1980.

[97] J. Uffink, "Bluff your way in the second law of thermodynamics," *Stud. Hist. Philos. Modern Phys.*, vol. 32, pp. 305–394, 2001.

[98] L. R. Volevich and B. P. Paneyakh, "Certain spaces of generalized functions and embedding theorems," *Russian Math. Surveys*, vol. 20, pp. 1–73, 1965.

[99] H. C. Von Baeyer, *Maxwell's Demon: Why Warmth Disperses and Time Passes.* New York, NY: Random House, 1998.

[100] J. C. Willems, "Dissipative dynamical systems, part I: General theory," *Arch. Rational Mech. Anal.*, vol. 45, pp. 321–351, 1972.

[101] J. C. Willems, "Qualitative behavior of interconnected systems," *Ann. Syst. Research*, vol. 3, pp. 61–80, 1973.

[102] J. C. Willems, "Consequences of a dissipation inequality in the theory of dynamical systems," in *Physical Structure in Systems Theory* (J. J. van Dixhoorn, ed.), pp. 193–218. New York, NY: Academic Press, 1974.

[103] J. Woodhouse, "An approach to the theoretical background of statistical energy analysis applied to structural vibration," *Journal of the Acoustical Society of America*, vol. 69, pp. 1695–1709, 1981.

[104] B. E. Ydstie and A. A. Alonso, "Process systems and passivity via the Clausius-Planck inequality," *Systems Control Lett.*, vol. 30, pp. 253–264, 1997.

[105] H. D. Zeh, *The Physical Basis of the Direction of Time.* New York, NY: Springer-Verlag, 1989.

[106] J. M. Ziman, *Models of Disorder.* Cambridge, England: Cambridge University Press, 1979.

Index

www.ingramcontent.com/pod-product-compliance
Ingram Content Group UK Ltd.
Pitfield, Milton Keynes, MK11 3LW, UK
UKHW031841161224
452263UK00003B/60

9 780691 123271